SIMULATIONS *A Handbook for Teachers*

Simulations

A HANDBOOK FOR TEACHERS

Ken Jones

Kogan Page, London/Nichols Publishing
Company, New York

First published in Great Britain in 1980 by
Kogan Page Limited, 120 Pentonville Road, London N1 9JN

First published in the United States of America in 1980 by
Nichols Publishing Company, PO Box 96, New York, NY 10024

Library of Congress Cataloging in Publication Data
Jones, Ken.
    Simulations.

    Bibliography: p.
    1 Education—Simulation methods—Handbooks, manuals, etc. I Title.
LB1029.S53J66 1980 371.3'97 80-14973
ISBN 0-89397-090-5

Printed in Great Britain by
Redwood Burn Limited
Trowbridge and Esher

ISBN 0 85038 319 6

Contents

Introduction

The aim of this book is to help teachers deal with the practical aspects of choosing and using simulations. Theory and jargon are reduced to the minimum and the intention is to clarify and de-mythologize the subject.

Although it is written for teachers who know little or nothing about simulations — and may have even picked up a few misleading ideas — it may also be useful for teachers already using simulations regularly, since even experts are sometimes unaware of simple ways of improving the presentation of a simulation and avoiding pitfalls.

Four themes recur throughout the book:

1. The problem of misleading terminology
2. The concept of reality
3. The concept of power
4. The search for wider objectives and improved techniques.

Although the book contains examples of actual simulations, these are not intended to be a list of 'best buys' — they are used simply to widen horizons and give an insight into simulation design. In this way the book will be a sort of 'do-it-yourself evaluation kit' for examining simulations, for selecting the ones that seem most appropriate for the teacher's objectives and the needs of the students, and for introducing the simulations in a relaxed and effective manner.

Chapters 3 and 4 constitute the main part of the book. 'Choosing Simulations' deals with the question of matching the objectives with the package of simulation materials and 'Using Simulations' goes into practical details, giving options and making recommendations. If the teacher has time to read only two chapters, these are the chapters to read.

The key chapter in the book, however, is Chapter 1, 'What's a Simulation?' Because of the misleading terminology of simulations there is a problem of unlearning as well as learning what a simulation is. Misleading words can be trip-wires. How not to get it wrong is as important as how to get it right. The criterion in this chapter is not a search for a 'correct' answer or definition, but a search for functional concepts which will help and not hinder the flow of a simulation.

Chapter 2, 'Design', contains examples of simulations — TENEMENT, STARPOWER, DART AVIATION LTD, and SPACE CRASH. It also looks into the author's own simulation workshop since it is of practical value to a teacher to be aware not only of what a designer puts into a simulation, but also the sort of things that are left out. Authors rarely explain why they have designed a simulation in a particular way and inexperienced adaptation by teachers can inadvertently reintroduce ideas which were discarded by the author because they did not work well.

Chapter 5, 'Assessment', moves into some controversial areas of research and assessment and recommends that the aims should be limited but the assessment should be thorough. It also proposes that, as well as using students to test simulations, the teacher should consider using simulations to test students.

Chapter 6, 'The Way Ahead', examines the possible development of simulations, particularly in the areas of business, secondary schools and the teaching of English as a foreign language.

Chapter 1
What's a simulation?

Why words matter

Terminology is the dragon at the simulation gate. Words convey
expectations; expectations influence behaviour; and if the words are
wrong the behaviour is likely to be wrong.

Many of the descriptive words in simulation literature have been
borrowed from other fields, particularly from gaming and the theatre.
Such words are: game, gaming, play, player, act and actor. Words
conjure up associated phrases: 'it's only a game', 'making a game
of it', 'play acting', 'acting about', 'playing around', 'acting it up',
and 'playing it for laughs'. Words are not isolated units.

Some authors see no harm in this, but probably most writers on
simulations are seriously concerned about the terminology. But
writers on games tend to regard simulations as a type of game, and
writers on drama in education seem to equate simulations with
role-play exercises.

The best example of the confusion is between simulations and
games. Some see the two words as interchangeable (Boardman,
1969). Others think that a simulation is a type of game, while
Tansey and Unwin (1968) define a game as being a type of
simulation. The double-barrelled 'simulation/game' is frequently
used, sometimes as an all-embracing category. On the other hand,
Bloomer (1973) makes a useful distinction by defining games and
simulations as being two distinct concepts and uses the phrase
'simulation/game' only to cover those which have the properties of
both. As Bloomer remarks, 'There can be few areas in which
semantic clarity is a more pressing need.'

If the teacher presents a simulation and the students try it out
expecting fun and games, they may be disappointed if it does not
live up to their expectations and decide to help it a little by injecting
a fun element. Similarly, if the teacher and students are used to
role-play and drama and informal drama and regard the simulation

9

as something theatrical, what occurs on the classroom floor is unlikely to be a genuine simulation.

Thus, the wrong words can lead to a simulation being inadvertently sabotaged, and the result will probably be far less satisfactory than if it had been a genuine game or a genuine informal drama. Since both teacher and students will think that the unsatisfactory experience was a simulation, it will not be surprising if they tell their colleagues that simulations are unsatisfactory.

Nor are the effects of misleading terminology limited to the people directly involved in a simulation; parents, librarians, school governors and administrators can all have the wrong impression. Even authors of simulation literature and researchers in simulations can be misled in ways which make a functional difference to what they do about simulations and how they assess them.

Consequently, words matter. Definitions and descriptions of simulations are not simply academic exercises or intellectual diversions.

From a functional point of view, definitions are far less valuable than descriptions, and descriptions are not as useful as observing what actually happens in a genuine simulation. It is also important to have the questions right — to know what to look for.

Inside a case study

Instead of thinking of a simulation as being like a game or an informal drama, it is more useful to think of it as being like a case study but with the participants on the inside, not on the outside.

In a case study the students sit on the outside and examine, with impartial detachment, a particular case including the documents related to the case, and they discuss among themselves what should or should not be done. They take an Olympian view and look at the whole picture, weighing things up and reaching a judgement rather like judges in a court case.

With a simulation the participants are on the inside, with the powers, duties and responsibilities for shaping events. Unlike a case study, it is not a static event. Action and interaction take place. The situation changes. Causes have effects and decisions have consequences. The participants are involved, they participate, they become absorbed in the interaction. It can be said that the participants *are* the simulation.

Another benefit of thinking of a simulation as being like a case study is that a case study is a more neutral concept than either a game or an informal drama, and this avoids the possible emotional associations of 'fun and games' or 'amateur dramatics'.

A simulation, like a case study, is essentially a group activity with a serious educational purpose. Like a case study it can be related to any subject from accountancy to zoology.

This argument does not imply that games cannot be serious or educational or that informal drama is always emotional and 'dramatic'. Even less does it suggest that simulations are somehow 'better' than games or drama. But, as will gradually emerge during this book, there are essential differences between simulations and other methods, and if simulations are to be likened to any other specific method then it is safer and more useful to liken them to case studies.

One highly significant feature of a simulation is the function of the teacher. Unlike most other educational methods, a simulation is not taught. The teacher is on the outside, not the inside; he or she does not participate in the interaction, has no powers or responsibilities for the decision-making, and is responsible only for the mechanics of the simulation — when it should begin and end, how the documents should be handed out, who's who, who sits where, and so on.

As a consequence of this change of function, the teacher is given a different name — usually the controller. This control refers to the mechanics, like the mechanics of looking after a car. The aim is to see that it runs smoothly, but it is the motorist — the participant — who decides which way to go.

Just as there is no teacher in a simulation, neither are there any pupils or students. They are participants. They can have any of hundreds of functions — explorer, judge, lawyer, journalist, statesman, soldier, spaceman, electrician, businessman, king, environmentalist, architect, editor, town-planner or even (if it is a simulation about education) lecturer, tutor, student or head-teacher.

Thus, a simulation is not teacher-student orientated. The participants have powers and responsibilities, and the controller has none of the attention and tension of teaching, but has a more relaxed function — well suited to observing what is going on.

Three parts

A simulation usually has three parts — briefing, action and de-briefing.

The briefing relates to the explanations given by the controller about the mechanics and procedures. Normally there is only one initial briefing in a simulation but it is possible in a multi-stage simulation to have mini-briefings inserted at specific points before some change in the activities.

Often there is a clear and sharp break between the briefing and the action itself. But in simulations where there are quite a few documents there may be little practical difference between reading a document during the briefing and reading it at the start of the action, when the participants are already divided into their respective groups or individual functions or roles. Some documents are read during the briefing, but are also re-read or referred to during the action itself.

Although it is not always possible to identify clearly where the briefing has ended and the action has begun, there is rarely any such problem in the de-briefing. Once the action has stopped, the participants revert to their normal functions as pupils or students and the controller becomes teacher again. The de-briefing is an inquest on what happened and a discussion of points arising.

At least, that is the theory. In practice, if the simulation is a powerful one, some participants may have difficulty in changing hats immediately. They will still continue to talk and react to some extent as if they were in the action. Re-entry is not always easy or instantaneous — not that this is necessarily a bad thing.

From this description it can be seen that the word simulation can be used to refer to the whole operation — briefing, action and de-briefing — but it can sometimes be used to refer specifically to the action part alone. If the distinction is relevant, the context will usually give the clue.

Similarly, the context should usually make it clear if there is any significant distinction being made between the word simulation as referring to an event, and simulation as referring to the materials on which the event is based.

If someone says 'She bought a simulation', 'He wrote a simulation' or 'They keep the simulations in the library', then it is clear that simulations are being referred to as packages of materials. But if a person asks 'Did you enjoy the simulation?' or 'Who was the king in the simulation?', then the word refers to an event.

It may seem unfortunate that the same word can be used for two different concepts, but there are plenty of precedents. A play can be written, read, or bought (materials), and a play can be acted, seen, and applauded (event). 'Case study' can refer to the materials or the activity based on the materials.

From the point of view of the teacher, it is better to think about, talk about and describe simulations as events rather than materials. The reason is that unlike a novel, poem or textbook, a package of simulation materials is incomplete. The best bit — the interaction — is left out.

Even people experienced in the use of simulations find it difficult, if not impossible, to judge the quality of a simulation simply from an inspection of the materials. With a good simulation there are often all sorts of clues and hints and opportunities for action that are not signposted in any obvious way. With a play at least the actors have specific lines and there are stage directions.

The danger of referring to a simulation as if it were the materials is that people will pick up the package and try to evaluate it. It is all too easy for the uninitiated to inspect the materials and then complain that such and such a point is not mentioned, or not stressed sufficiently, or not driven home, or to make a general complaint that the whole thing is too complicated and too difficult. It is frequently the case that such judgements are mistaken.

Starting an event

The best way — and some people would say the only way — to understand what any specific simulation involves is to participate in it. But although no written description is a substitute for participation, it is possible to take imaginary journeys into the thoughts and feelings of a participant — particularly in the briefing and reading-in stage before the action starts.

So . . .

You are 16 years of age (or an adult if you like) sitting down with other potential participants awaiting the briefing for your first simulation. You have heard of the word, but you do not know what it means in an educational context. Your teacher, however, has personally participated in several simulations and has been controller in half a dozen others. Moreover, the teacher knows about the problem of terminology and might well start the briefing by explaining why it is useful to think and talk in terms such as 'participate' and 'participant', rather than say and think 'act' or 'play' or 'actor' or 'player'.

The teacher explains that you do not have to act a part. You remain yourselves with your own personalities, but with a different function.

You have a job to do and you try to do it in the best way you can in the circumstances in which you find yourselves.

For example, says the teacher, if in a simulation you had the role of king, you would not have to pretend that you were any particular king that came to mind — Henry VIII or Charles I. You would not have to pretend at all, since you would be a king by your function, you would have the powers and duties and responsibilities and problems of kingship. If you were a reporter in a simulation, you would not say to yourself: 'Now I'll pretend I'm that reporter I saw on television last night and I'll act like he acted.' You would actually be a reporter, not a pretend reporter, because it would be your job to report on something and you would be responsible for what you did. So you could not say afterwards: 'Well, I know it was silly, but I was just pretending to be a reporter.' In a simulation, if you are a reporter you are a reporter, not an actor or a mimic or a comedian.

The teacher then says that in a simulation there is no teacher, since a simulation is not taught; the teacher becomes the controller in charge of the mechanics of the simulation. This means, says the teacher, that you are responsible. You have to make your own decisions. Do not ask me about any policy question — you have to sink or swim on your own. You have the power and it is real power. I am not going to interfere; providing you stay within your function you can do whatever seems right or best and I am not going to give hints or advice or any other little signs of encouragement or discouragement. So when the action starts, forget about me altogether.

In giving you this sort of explanation, the teacher may or may not be helped by the simulation materials themselves, particularly the controller's notes. Usually, the notes say nothing about simulations as simulations and concentrate entirely on the mechanics, plus follow-up questions for the de-briefing. Nor are there many simulations which provide notes for participants (as distinct from role cards) which back up the controller's briefing.

So rather than follow an imaginary briefing, suppose that the simulation is the author's RADIO COVINGHAM, and the controller says, 'Now, to help you understand what it's all about, here are the notes for participants', and hands you this document (see opposite).

The value of having notes for participants is that they should cover all the main points, thus making the teacher's job that much easier. Without such notes for participants there is a danger that the teacher may leave out some important point during the briefing

NOTES FOR PARTICIPANTS   *RADIO COVINGHAM*

What it's about

You produce and broadcast a ten minute programme on Radio Covingham called *News and Views at 7.* You'll find most of the information about how to do it in the Station Manager's Memo. Read it very carefully. When it starts you'll get a pile of listeners' letters and handouts. These came in the morning mail. But the news items will come in one by one, gradually, right up to the time you go on the air. The Controller will brief you, and tell you the exact time you go on the air.

What you can do, and what you can't do

You can rewrite the material, but you can't invent news.

You can interview members of your own team in various roles, say as someone who wrote a letter, or is connected with a handout or the news items, but only in connection with the material.

You can, if you have a tape recorder, use it to see what an item sounds like, but you cannot use taped inserts in the broadcast itself. The broadcast itself should be recorded on tape, but it must be completely live, with no pre-recordings fed into it.

If you have time you can have a rehearsal, but this will not stop news flowing in.

Advice

Ten minutes is longer than you think! So aim for at least 2 interviews, preferably 3 or 4, and possibly 5 or 6.

Before the simulation begins, decide on how you are going to handle the material — perhaps with three groups dealing with letters, handouts and news.

If you are the producer, don't get bogged down in individual items, particularly if you have a large team. Make sure that someone is in charge of the timing — perhaps using hand signals for slowing down or speeding up, or putting up fingers to show the number of seconds or minutes remaining.

Before you begin, decide on the general shape of the programme. Should news be bunched together at the beginning, in the middle, or at the end, or should it be scattered throughout the programme? Will you have news headlines at the beginning? Will you have news headlines at the end? Will you have a special section or spot on a particular subject — such as entertainment, business and industry, etc? But you'll have to wait until you get the materials to decide the subjects. And remember the programme must contain 'views' — this means you can comment on items.

If you've finished your particular job then don't just sit around — ask the producer what else you can do.

PLEASE DON'T WRITE ON THE MATERIALS. THEY WILL BE NEEDED AGAIN.

Nine Graded Simulations
No. 3 Radio Covingham

and have to interrupt the action later on and say, 'Sorry, I should have told you that . . .'

It can also be useful if the notes for participants help to draw the line between what can and cannot be done in a simulation. In an informal drama about a local radio news and views programme, it would be an excellent thing if the participants invented news items and produced interviews about imaginary events. But in a genuine simulation the journalists are journalists not authors of fiction. As the notes point out, 'You can rewrite the material, but you can't invent news.'

Unless the distinction is clearly made between what can and cannot be done in a simulation, then even the best simulations can run away.

The notes for participants also refer to the station manager's memo. This could be regarded as part of the briefing even though it may be handed out after the participants have moved to their areas, ready to receive the listeners' letters, news items, etc. So a description of the briefing for RADIO COVINGHAM really needs to include this document (see opposite).

One of the main reasons for giving the information in the form of a separate document is to help the mental process of changing from student function to journalist function. This sort of device is not all that unusual in simulations — it adds a rouch of realism, it evokes an atmosphere, it conveys information indirectly, and it places the participants on the inside.

This example gives some idea of what a briefing might be like. But the problem with describing simulations is that the vital part, the interaction, is left out. If a simulation were a novel or a poem, each of which is complete in itself, there would be no such problem.

So a description of any simulation requires a good deal from the imagination, and what may be imagined may not correspond very closely with the sort of thing that happens in practice when that simulation is used. Even a description of a briefing is cold. There is no substitute for being a participant, for actually knowing that you had better pay careful attention to the documents since you are going to be involved. It is your responsibility, and you are not going to receive any help. If you are not involved, then you can read the documents related to a briefing in a detached and objective way. But it is quite a different feeling when you know that you are going to be sitting in the hot seat. A simulation is action, not a cerebral exercise.

RADIO
COVINGHAM
MEMO

To: Production staff of 'News and Views at 7'.

From: Station Manager, Radio Covingham

     I've received a rocket from the national network about over-running. Last Friday we were cut off in mid-sentence so that the network could come in at precisely 19.10.

     'News and Views at 7' is a ten minute programme. That does not mean ten minutes and ten seconds, or even ten minutes and one second. So please keep an eye on the clock.

     We can, of course, under-run a bit, even by as much as half a minute, which can be filled in with a continuity announcement or trailer. But aim to come out somewhere within the last ten seconds before the national network comes in. This can be done if you check the timing during rehearsal, and if you can have one or two optional items near the end of the programme which can be broadcast or not according to the time available.

     Reporters - excellent work on the plane crash last week. We beat the news agencies by more than half an hour. However, do remember that you should try to avoid asking questions in interviews which can be answered 'yes' or 'no'. Begin the questions with letter 'H' or 'W' - how? why? what? And follow up interesting answers. Listen to what the person is saying.

     'News and Views at 7' now has a wide audience in the area, and D.L.R.(N) made complimentary noises last week about your output.   Keep up the good work.

Colin L Taussig

Nine Graded Simulations
No. 3 Radio Covingham

Definition dangers

Most books on simulations start with a definition. But, as already indicated, simulations are frequently victims of misleading terminology.

Take, for example, the word 'simulation'. Suppose that the Ruritanian cultural attaché visits a school in Britain and the headmaster says 'We use quite a lot of simulations in our sixth form.' 'Simulations' says the attaché, nodding politely, and deciding to look up the definition of the word later. On returning to the embassy the attaché picks up a dictionary, looks up simulation, and discovers that it means to feign and counterfeit. For example, *Chambers Twentieth Century Dictionary* puts it like this:

simulate, *sim'ū-lāt, v.t.* to feign: to have or assume a false appearance of: to mimic.—*adj.* feigned. — *adj.* **sim'ulant,** simulating: (*biol.*) mimicking. — *n.* a simulator. — *adj.* **sim'ular,** counterfeit: feigned.—*n.* a simulator.—*n.* **simulā'-tion.**—*adj.* **sim'ulātive.**—*n.* **sim'ulātor,** one who or that which simulates.—*adj.* **sim'ulatory.** [L. *simulāre, -ātum*; cf. **similar, simultaneous.**]

After reading this definition, the Ruritanian attaché looks up other words — feign, false, mimic — and finds a host of similar words: dissemble, counterfeit, invent, pretend, wrong, deceptive, untruthful, untrue, artificial, one who performs a ludicrous imitation of others' speech and gestures, an unsuccessful imitation, a mime actor.

This is amazing, thinks the attaché. All these words paint a consistent picture of something highly undesirable in education — false, deceptive, artificial, ludicrous, and so on — thank goodness simulations are not used in Ruritanian schools. Simulations must be one of those gimmicks of which the British and Americans are so fond.

This dictionary-based conclusion may not be too far removed from the impression which the word simulation gives to people who know more about British education than the attaché — parents, administrators, college governors, librarians or any teacher who asks the question 'What's a simulation?'

It would be better if simulations were known by some neutral or

invented word — teggits, or blarks, or named after a pioneer
(Twelkers), or (something more descriptive of what actually
happens) realsits (Jones, 1980a). Unfortunately, the word simulation
has been accepted currency for too long to be changed, but any
search for a new word, no matter how unsuccessful, at least draws
people's attention to the defects of the existing word.

It would be encouraging if authors of simulation literature repudiated
the dictionary definition, but this has not been the case. On the
contrary, definition after definition stresses the concepts of unreality
and artificiality. Here are three fairly well-known definitions:

> *Garvey (1971)* A simulation is the all-inclusive term which contains
> those activities which produce artificial environments or which provide
> artificial experiences for the participants in the activity.

> *Guetzkow (1963)* A simulation is an operational representation
> of the central features of reality.

> *Taylor and Carter (1970)* Delineates a range of dynamic
> representations which employ substitute elements to replace
> real-world components.

The words 'real' and 'reality' occur in two of these definitions, but
only to exclude it. It would seem from these definitions that a
simulation is about as real as a photograph of a person compared
with the real person. If one did not already know what a simulation
was, then any of the following items might suggest themselves as
being included in some or all of the definitions:

- — A type of television programme with artificial environments.
- — A musical score being performed.
- — The dynamic use of maps during fieldwork.
- — The written account of an action involving the central features
  of reality.

This speculation is not fanciful. Tansey and Unwin (1968) mention
maps, graphs and circuit diagrams as examples of simulations.

If the above definitions were related to what happened in a
simulation, then the participant who emerged from a useful
simulation might be expected to say something like 'That was a
good artificial experience using substitute elements.' To which the
teacher might respond 'I appreciated your marvellous use of artificial
talk and representational behaviour.'

This is the opposite of what happens after a good simulation. If, for
example, the simulation includes an interview, then the participant
may say, during the de-briefing: 'That was a *real* interview, not a
pretend one. My palms really did sweat a bit. I wasn't acting at all,
it wasn't a game, and we were not playing about. That was real

life, far more real than anything else we've done in class this year.'

A much better definition of a simulation is given by Twelker and Layden (1973) who say that simulations are simplified reality. The word 'simplified' is not altogether helpful since it suggests that simulations cannot be complicated. They can. There are many long and involved simulations which cover a mass of detail and have considerable complications. But if 'simplified' means that it is neither desirable nor possible to 'put in every detail' in a simulation, then this is quite true.

If a short definition really is necessary, perhaps it might be: 'Simulations are reality.' From the teacher's point of view at least it errs on the right side, the side of function, the side of presenting good simulations and avoiding poor ones.

Probably no definition of a simulation can be entirely satisfactory since it is trying to write small what should be writ large. It is concepts that matter; the concepts of reality and power are two key issues in understanding simulations and each deserves a prominent section on its own. In examining the concept of reality, it will be seen that it can be useful for the teacher to take some of the definitions of simulations and stand them on their heads.

Reality

The word 'reality' is chameleon-like, sometimes taking on one colour, sometimes another. And what is 'unreality' — is it 'artificial' or 'fiction' or 'fantasy'?

In the definitions already discussed, a distinction was made between reality and representation, and what happens during a simulation was labelled as representational, compared with the real and non-representational world existing outside the place where the simulation occurs.

It is useful and functional to assume the opposite of this. Paradoxical though it may seem from the point of view of both controller and participants, reality exists — genuine and actual — in the essence of a simulation, the interaction, while the representational part is outside the simulation area.

It works like this. The participant has a real function — to interview, to report, to analyse, to speak, to think, to negotiate, to understand, to communicate, and so forth. So when the participants speak in the people's assembly the speeches are perfectly real and have real consequences. They may be amateurish, naive, lacking in persuasion, but they are still real. If the simulation contains a role for a journalist, then the journalist's report of the proceedings would be real, not a substitution or a representational element, or an artificial facsimile. A real person is writing a real report about a real event.

In a business simulation, the business may be conducted by people who have never been in business in their lives, but their thinking and talking and decision-making are absolutely genuine — it is not like a play, a pretence, or a game. The participants are not imitating businessmen, they *are* businessmen because they are functioning as businessmen.

The representational part is outside the area, outside the classroom. The corridors outside the door are the corridors of the school or college, not the corridors of power. There is no factory down the road making the products which are causing so much discussion among the businessmen. These elements are representational, artificial, substitute and imaginary — they represent an appearance of reality.

This distinction between the reality inside and the representation outside applies even if the outside is seemingly real. Suppose, for example, that the simulation is taking place in the Pentagon in Washington and the participants are themselves high-level diplomats and defence department officials, and that the simulation deals with the real world as it exists at that precise moment. This makes no difference. There is no contact with the outside world. The 'hot line' in the simulation leads to the controller or to another participant, and not to the Kremlin. By definition, and in simulation terms, the outside is still representational.

This distinction between reality inside and representation outside is not just a description. It is functional and helps the controller and the participants to draw the line between what can be done and what cannot (or should not) be done in a simulation.

The government of Blueland can order the mobilization of the army. The order is legitimate since that is within the reality of the simulation. But the effect which the order has on the representational world outside is something quite different and comes under the authority of the controller or the control team. It may be that control decides it is plausible to assume that the mobilization proceeds as planned and envisaged, with no untoward incidents or delays. On the other hand, the situation might arise where the order is disobeyed or results in riots. Again, the participants have no control over this — it is outside the room.

One significance of the distinction is that the participants cannot increase their powers; they cannot become supermen or magicians

21

or gods, shaping the world according to their whims or desires.

The dividing line exists, or should exist, in small-scale as well as in large-scale simulations. If the simulation is about a local community's discussion on where to site a footbridge over a stream, at some point a participant might think it a good idea to invent a 'fact' in order to strenthen his argument. If he gets away with it and if the other participants accept that his allegation constitutes a simulation 'fact', then they may start to play the same game. They too will dabble in authorship, inventing this and that.

Unfortunately, in most simulations, neither the materials themselves nor the controller's briefing contains any warning about the danger of mistaking what is real and can be done, and what is representational and cannot be done. The probable reason is that most authors are unaware of the true causes of what is usually referred to as the 'Armageddon syndrome' — an escalation to disaster. This is usually thought of as endemic to large foreign affairs simulations, but a small-scale simulation can be blown up just as easily.

One recommended device is for the controller to publicize a final session, but not to hold it — and to end the simulation earlier. This fails to tackle the cause of the problem and has the danger that the students may become aware of the device, and thus bring Armageddon forward to the penultimate session.

The sort of thing that happens is that invented facts become increasingly absurd. Imagination flows recklessly to the heads of the participants. Instead of talking about negotiation and diplomacy, they become inventors. 'We've mobilized our army and put one million troops on the frontier,' claims one prime minister. The rival prime minister counters with 'And we have mobilized all our armed forces — army, navy and air force — and there are now two million of our troops on your border.' From this, it is a short step to 'We've won because our paratroopers have attacked your capital and we've captured you' and 'No, you haven't because we've captured your paratroopers and now we've just dropped nuclear bombs all over your country, and you are all dead.' 'No, we're not, because . . .'

Thus it can be seen that a simulation depends on reality, otherwise it ceases to be a simulation. A diplomat must be a real diplomat, not a magical diplomat. A participant who is a local councillor is real only so long as he deals with simulation facts and does not start taking on the function of local wizard. Not only are simulations ruined in this way, but the reputation of simulations is also damaged, since neither participants nor teacher may realize what has happened.

The initial reaction might be, 'That was all right. Lots of enjoyment and participation, and some tremendous creative imagination.' But some participants would be left with a sour taste, feeling that their earlier efforts had been sabotaged by cheating. And the teacher might tell colleagues in the staffroom, 'They seemed to like it, and there was some good acting and plenty of creative ability, but I'm not sure that they really learned anything from it. I don't think we'll try another simulation just yet.' And no one will realize that the simulation lies dead on the classroom floor. What took place was not a simulation, but a pseudo-simulation.

If, however, the teacher or the materials draw the attention of the participants to what can and cannot be done, then several benefits result, not just a safeguard against sabotage.

One is to remove any misconceptions that behaviour within a simulation is, or should be, unreal. There is no excuse, no opting out, no saying 'Well, I was simulating, wasn't I?' The participants are responsible for their own behaviour. Unlike some informal dramas or role-play exercises where the participants are told 'You are angry', 'You are obstinate', 'You are weak', a genuine simulation does not try to control behaviour — behaviour depends on the participant and is real, not assumed.

Another benefit is that fewer participants will wish to try to 'cross the line' if it is pointed out that the outside world is representational and is within the authority of the controller, not themselves. Issuing an order has uncertain consequences and the participants may well feel that it is better to keep the situation within their own control by negotiation, discussion, debate, and so on.

There are some other aspects of the concept of reality which can help the simulation user. One is to avoid thinking that reality means detail. The teacher may be tempted to think, 'In a game, like chess, there is little or no reality — a real bishop does not travel diagonally, nor does a real king move one square at a time. But, in a simulation, there are supposed to be lots of real-life details and the more details the better.'

But as every simulation designer knows, detail can clog and confuse and kill a simulation. It depends, of course, on the particular simulation, but with every simulation there is some point where the addition of detail adds unnecessary complexity, and may unbalance the simulation with the introduction of more roles, more organizations, more documents, more small print. The scenario becomes unwieldy, participants have passive roles or part-time roles and sit around with little to do, and the simulation becomes unworkable.

In any case, if the aim of the simulation is to show some economic, political or social model in a simple form, then additional details will make it less like a real model and more like a specific example.

For the same reason it is not helpful to think of reality as necessarily excluding fantasy. A simulation designer may decide to place a simulation one thousand years into the future, or locate the simulation on another galaxy, in order to manipulate the environment so as to clarify 'reality' by removing confusing or historic associations. Existing institutions contain features which are irrelevant and non-essential and fantasy is a useful device for examining the essentials. Fantasy can be used to challenge the participants to ask basic or ultimate questions.

Reality can be a useful concept in the design and presentation of documents within the simulation, although in this case 'realism' might be a better word. Documents which look like the real thing are usually preferred by the participants to the same items of information presented in a typed or printed format on identical sheets of paper. If a citizen's letter or a chairman's memo or a foreign ministry document looks something like the intended example, then even adults appreciate the touch of realism.

Power

Power is an essential concept. Simulations are owned by the participants. It is they who have the powers and the duties in the fields of policy and decision-making. If the controller does intervene in these areas, or if the simulation itself contains chance elements (dice, cards or random numbers), or if the role cards seek to imprint personalities, then participant ownership is reduced and the simulation becomes that much less of a simulation.

Participant ownership is vital. The most memorable experiences which students can derive from simulations are connected with power. They must have authority. For lonely, shy and introverted students the experience of being a person of importance can be traumatic, a vision of a new world, of excitement, of growing confidence, and a feeling of being wanted and respected. Such experiences can sometimes be remembered with the utmost clarity for years afterwards. Ownership encourages communication; power and responsibility lead to talk; this facilitates thinking.

If the power structure is group power — say that of a committee with no individual roles within the committee — then this allows a shy individual to shelter in inactivity. In such a simulation it may be difficult to encourage the non-talkers to talk. One way in which a simulation designer can deal with this is to raise the stakes. If the penalty of failure is death, as in a survival simulation, even a shy person may utter a cry of warning or put forward a good idea. Another technique is to use highly controversial issues. Mouths can be opened by a sense of injustice or outrage. If, on the other hand, the power rests in individuals rather than in groups, the opportunities for communication and language increase, but so do the dangers to the shy pupil who does not wish to express himself.

Fortunately, in a good simulation, the shy participants have no time to sit and raise defensive battlements around their egos. They have a job to do. They are involved in a motivational situation. They have duties and responsibilities and are too busy and too conscientious to worry unduly about making fools of themselves. After a few anxious moments at the start of their first simulation, they are prodded or pushed by the other participants in the interaction. They are interviewed or consulted or action is demanded. So the shy students, finding that the contact is one of equality, respect and duty, say or do something according to their functions.

With power comes courage. They say to themselves, 'I have to do this; it is my job.' This is an attitude of action and is in sharp contrast to the near paralysis a shy person may sometimes feel in a teacher/student situation.

In this way, a simulation provides a parallel with what happens in the outside world. Most people, for example, seem to think that newspaper reporters are pushing, persistent, aggressive and intrusive people. In fact, most reporters probably do not have this personality. Most of them are probably introverts, having entered journalism because they are good at writing. They push and persist because it is their function, duty and responsibility to get a story. But doing so also enlarges their experiences and makes them more confident, and results in a facility for finding effective ways — and not always aggressive ways — of getting the news. The same sort of job experience occurs elsewhere — not least in the field of education. Dedication to a job brings bravery in its execution.

One consequence of power in a simulation is a possible or probable shaking up of the pecking order among the students. This occurs since the motives and incentives for action, and the opportunities for action, are very different from what they are in teacher-student orientated lessons. It is a common experience of teachers using simulations to find that their expectations of student behaviour go wildly astray. Little Sally can blossom forth in a way that can quite astonish the teacher, her colleagues, and even herself. John, on the other hand, who usually has plenty of attention and esteem by shouting out the answer first, finds that there is no 'answer' to shout out, no 'shouting' at all, and that success may require a reasoned discussion rather than an instant answer.

Responsibility

The power and ownership of a simulation relates, of course, to the action part; and this power and ownership entail duties and responsibilities. It is not the irresponsible power of madmen; it is the power that is related to function, to office, to doing a job.

Acceptance or evasion of these responsibilities is the key to the success or failure of a simulation. It is the responsibility of the designer and also the controller to see that the participants accept the responsibility as well as the power. This does not mean that the controller keeps interfering and saying, 'You're not accepting your responsibilities.' It means that the duties and responsibilities should have been spelt out in the briefing or incorporated in the documents or both. It also means that the procedures and functions should have been adequately explained.

Suppose that the participants are members of parliament in Blueland and that part of the simulation is a debate on a controversial issue in the people's assembly. It is not enough for the author to provide material for the debate; the procedure should also be made reasonably explicit.

This means deciding who is in charge — a speaker or chairman — and what are the formal powers. The chairman and the MPs must be told how things should be done. Do speakers have to stand up to address the assembly? Do they have to address the chair? How do members indicate that they wish to speak? Can speeches be interrupted by other members? How does the chairman keep order? Can the chairman suspend the session and if so, how? Can the chairman ask a member to withdraw from the chamber? If a member proposes a motion, is that member allowed to reply to the debate? Are amendments permitted? How is a vote taken? What majority is required? If any of these sort of problems might crop up, it is better that they should be dealt with before the assembly meets.

Responsibility implies authority and if the participants are uncertain about where authority lies, one of two things will happen. Either the participants will halt the simulation, stop being members of Blueland parliament, and appeal directly to the controller for information and advice, or else they will struggle on and invent their own rules (or lack of them). In the first case, the flow of the simulation will be badly interrupted and it may take some time for the participants to resume thinking like members of parliament. In the second case, the result can vary from the mildly unsatisfactory to the completely disastrous.

If the debate in the people's assembly starts getting out of control and degenerating into a mindless shouting match, the controller may feel obliged to intervene anyway. The controller, who has been sitting quietly at the back of the room, observes that three or four members of parliament are on their feet shouting angrily at each other across the chamber, while the chairman is sitting silently, looking cowed and casting apprehensive glances in the direction of the controller.

This sort of problem will be dealt with in Chapter 4 on using simulations but, briefly, the best technique is to try to handle it within the simulation. It is not a good idea for the controller to march on to the floor of the chamber and say, 'Now then, cool it, let's have some order in the debate, otherwise we'll have to . . .' This stops the simulation dead in its tracks. Also it serves notice on the participants that they had better behave themselves in the future, otherwise the controller (or rather the teacher) will intervene. Such interventions diminish participant responsibility. The controller has taken over responsibility for law and order. What is happening is not a genuine simulation, but a quasi-simulation, approaching a teacher-controlled exercise.

What the controller should do is to send (take) a note to the chairman of the assembly saying something like 'Please suspend session for immediate and urgent consultations' and sign it 'President Ricardo' or 'King Andrei VI' or some other appropriate authority. This retains participant responsibility, and gives the controller and the participants a chance to establish some procedural rules of debate, which should have been done in the first place.

Secondary school discipline

Some teachers in secondary schools are afraid of trying simulations because they believe they could easily get out of hand, with participants playing about and jumping around enjoying a brief moment of freedom from teacher control. The word simulation itself, and the fact that many simulations are referred to as games, may reinforce the apprehensions. If a colleague says 'The teacher's not in charge of a simulation — the pupils are given the power to do what they want', the teacher may firmly close the classroom door on any simulation.

This is not an unreasonable attitude. No teacher wishes to be responsible for the behaviour of the class without the authority to enforce it. Consequently, the teacher may say 'I can see that simulations may work well with mature, disciplined and highly motivated groups, but they would be quite unsuitable for the pupils I'm in charge of.'

Even supposing that the teacher was persuaded to try a simulation, there would still be anxieties and uncertainties in the presentation of the simulation which would surely be picked up by the sensitive antennae of the pupils who might try to take advantage. Since the apprehensions were caused in the first place by a failure to appraise the nature of a simulation correctly, the briefing would presumably convey to the pupils some of the misconceptions about reality, power and responsibility.

It may be, of course, that the simulation is sufficiently strong, explicit and interesting to avoid disaster, or even mishap, but the risk remains. For example, the teacher, believing the simulation to be some sort of game, may give it a low priority spot on the timetable — a Friday afternoon or immediately after an examination. This may encourage the pupils in their belief that they are about to have a bit of relaxation, and they may oblige by playing it for laughs, enjoying some good gaming fun, and generally acting up. Both teacher and participants may assume that this is normal behaviour in a simulation. The event could end amicably, but the probable conclusion would be that simulations can be fun, but have little educational benefit. Again we have a situation where mistaken ideas prevent a genuine simulation from occurring, yet everyone assumes that it has occurred.

Suppose, however, there is a threat of misbehaviour. Perhaps there are signs that pupil A is about to lose self-control, pupil B is out to play the fool or pupils C and D wish to opt out. If this occurs during the briefing or de-briefing there is no special problem, because it is the same as in any other teacher-pupil situation; the teacher is in charge of discipline and has the authority to deal with misbehaviour.

The problem, therefore, can occur only in the interaction part, as this is the part of the simulation over which the teacher has no disciplinary control. But, since this is the case, the threat of misbehaviour is directed at the other participants, and they will usually deal with the threat themselves, either within the simulation itself (the chairman of the people's assembly calls for order) or else by bluntly telling the offender to stop messing about. It is like a casual game of football among youths in a park, where a goalkeeper gets annoyed or fed up and decides to lie down in the goal mouth. There are various things the controller can do to help — usually techniques for finding out what is really causing the trouble — and these points will be examined in Chapter 4.

But if the misbehaviour involves more than one or two participants and affects a whole group, there is no question of stopping the simulation — there is no simulation to stop. The participants have abandoned it and have in effect sacked themselves from their jobs. Therefore the teacher need have no compunction about intervening.

An analysis of what is involved reveals that the fears result from a failure to understand correctly what a simulation is. Not only does the failure cause unreal fears, but it also obscures the advantages. Pupil behaviour is usually much better in a simulation than in ordinary lessons. Pupil authority brings pupil responsibility, and this in turn leads to self-discipline. As will be suggested in Chapter 3, a teacher could use simulations with the sole aim of improving discipline.

Chapter 2

Design

Creativity

There is a story about a sculptor who had just finished making a stone elephant. He was asked, 'How did you do it?' He replied, 'I chipped away those bits that did not look like an elephant.'

Simulation design can be rather like this. Most people think of creative work as building, but it might be just as useful to think of it as chipping away. The simulation designer starts with a thousand potential simulations and ends up with one.

Creation involves closing options. Should the simulation be closed or open-ended? Should it be on a national or a local level? Should there be group roles or individual profiles? As each of these questions is decided, a great many possible simulations have been chipped away.

When a teacher inspects a package of simulation material, he or she may think that the author started off with a nice clear objective and the simulation was arrived at by some sort of logical deduction. What the teacher does not see is the author's wastepaper basket.

Authors must speak for themselves on this question, but I have not met one who regards creative writing as a simple matter of moving from a starting point, which is an objective, to a conclusion, which is the finished product, rather like a motorist plotting his journey on a map. Creative thought processes seem particularly difficult to recall afterwards — it is not like remembering what one bought at the supermarket. It involves a great deal of appraisal, discarding, selecting, and altering, and sometimes changing things around completely because of some new idea.

It is difficult to describe what is meant by design in a simulation. Some simulations have style, elegance, wit, and are not unnecessarily complex. A small number have a touch of genius, bringing a gasp of admiration for their ingenuity or profundity. A well-designed simulation has balance, the parts fit together well, and there is sometimes a degree of ambiguity, allowing different assessments of its meaning and impact.

With simulation writing, authorship is not usually enough; there has to be a testing of the product. Unlike writing a novel or a poem, the author cannot simply submit it to a publisher or ask for an expert opinion — it has to be tried out — and it is unusual for a simulation to be right first time. Often it needs considerable alteration and re-testing.

Consequently, one of the most important requirements for producing simulations is time. The average author requires about 100 hours to design a simulation, including the time required to test it out, alter it, and test it out again.

With 100 hours as the average, it is not surprising that many teachers, who think it might be useful to try designing a simulation for their students, find the effort and time required is not cost-effective and that there are more rewarding ways of helping students.

Nevertheless, every now and again a teacher perseveres with simulation design and the great majority of published simulations have been produced this way. A large percentage of this majority have the objective of giving information, insight and understanding about the specific subject taught by the teacher. Thus, a lecturer, class teacher or instructor involved in biology, geography, hospital management, advertising or sociology will design a simulation with the same aim, more or less, as if it had been a case study, or textbook, or guided exercise. Usually, the author is satisfied with writing one or perhaps two simulations.

In addition there is the small minority of simulation designers who specialize in simulations as such. They would think more highly of a good simulation about a poor subject than a poor simulation about a good subject. They tend to be the experts and the professionals, who read the simulation literature, who attend conferences on simulations, and who are members of societies dealing with simulations. Generally they manage to have their names on their products, either on the back of the pack in small print or, if they are more fortunate, featured prominently on the front cover. As with novelists, authors of simulations have their own styles, and some of them have a very distinctive style, so it is worth the teacher making a mental note of the name of the author. It must be added that many simulations do not give the name of the author. This may be because the simulation has been altered by many hands over the years, or because the simulation has been written by a committee. From the teacher's point of view it would be beneficial if all publishers made a point of identifying their authors.

Like everything else, appreciation of simulation design requires experience. The following four simulations serve to illustrate the diversity of design. Each has about it some special feature which, although not making it a typical example, makes it an interesting case to study.

TENEMENT

TENEMENT is one of the best known of British simulations and was written by some of the staff of the housing charity, Shelter. (The authors are not named in the simulation materials.) It concerns the problem of families living in a multi-occupied house in a large city. There are 14 roles: seven tenants, six agencies, and one landlord. 'd. Each has a role card of about 500 words. There are no other documents except for the controller's notes and some chance cards which the controller hands out if and when he thinks they might be useful.

It is intended for young people of 14 and upwards and the stated purpose is, 'to make young people aware of the difficulties and frustrations of living in such a situation, and to point to ways in which some of these difficulties could be solved by the introduction of agencies concerned with such problems'.

Looked at from the point of view of design, TENEMENT is particularly interesting as an example of excellent and emotive subject matter linked with a design which is weak on participation. The main problem is that although the tenants can visit whichever agency they think might be most useful in their individual circumstances, the participants who are in charge of the agencies might be overworked or have nothing at all to do. The participants in charge of the Citizens' Advice Bureau may sit at their table throughout the simulation twiddling their thumbs if the tenants decide to home in on those agencies which seem to have more relevance. The Department of Employment participants are also likely to have a long wait. They have only one potential applicant, Mr Johnston, as he is the only unemployed person. However, Mr Johnston and his family are the only tenants facing immediate eviction, so he may be too busy seeking help from the Rent Tribunal or the Housing Department ever to get round to visiting the Department of Employment.

The controller's notes are of little help since they do not mention the participation problem. They contain about 1500 words — half of which consist of suggested questions for the de-briefing.

Suppose that two or three of the agencies are sitting around feeling restless with discontent or bored with inaction. Should the controller refrain from interfering and hope that things sort themselves out, or should he walk over to the agencies and crack jokes about being unemployed? Should he search through the chance cards and, finding 'You have been made redundant', give it to the participant at the tail of the queue at the Housing Department?

Some of the chance cards might themselves cause bottlenecks. There is one for the landlord saying that a property developer has offered to buy the house, but only if there are no tenants. And if the landlord decides to try to evict the other six tenants, they will probably join Mr Johnston in a queue, or else all the tenants will start to negotiate with the landlord personally which would mean that all the agencies would be without customers.

Nor is poor participation the only design problem. What are the simulation facts? The controller's notes say nothing about this. Yet the Rent Tribunal's role card says, 'Remember, it is important to get all the facts. The tenant and the landlord may tell different stories, and the Tribunal will have to decide which one is nearer the truth.' But there are no authoritative documents or facts for the Rent Tribunal as the role cards are the only source of information. But lies are anticipated and the Williams family, for example, are instructed on their role card to conceal the truth about being evicted from their last council home and about the fact that they still owe rent. The door is open for lying and cheating, allegations and counter-allegations, and what may start as a simulation may well end up as an informal drama or a general free-for-all.

One of Shelter's education officers said that on one occasion the tenants tried to 'murder' their landlord and the Shelter official had to be called in to act as policeman to try to control them. On another occasion the old-age pensioner got down on his knees and proposed to the unmarried mother in order to share her flat, and one tenant decided to emigrate to Australia (Jones, 1974b).

It is not recorded what the participants at the agencies were doing while the tenants were having fun and games, but it seems clear that some of the participants were turning their imaginations to good account.

This demonstrates not only the disadvantages of a poorly designed simulation, but also the strength of a simulation which evokes sympathy and concern and which the participants may decide to rescue by transforming it into an informal drama. Unfortunately, not many simulations are as evocative about injustice and poverty as TENEMENT. If they are poorly designed, then any rescue attempt may be in the form of playing it for laughs.

STARPOWER

Garry Shirts is an American author of several highly ingenious and evocative simulations, including STARPOWER, which is the world's most used simulation. It too is about poverty and power, but in both content and design it is very different from TENEMENT.

STARPOWER has what is known as a hidden agenda. Things are not what they seem to the participants. Ostensibly it is a sort of trading game. The rules say that the primary goal is to become wealthy but that the participants can decide how to do this. Usually they decide in favour of competition. They are divided into three groups — squares, circles and triangles — and each participant takes his initial wealth from a grab bag. What is not realized by the participants is that most of the squares start with greater wealth than the circles, and that most of the circles start with greater wealth than the triangles.

After a few trading sessions the controller intervenes to identify the groups according to their wealth. To be a square requires a specific amount of wealth, so a small number of people who were originally squares are demoted to circles, and one or two circles are promoted to the top group, the squares. Similarly, a few promotions and demotions occur among the circles and the triangles. It looks as though the ups and downs were the result of trading skill, whereas in fact it was inevitable from the start because of the handout of original wealth.

Already the behaviour of the participants has changed. Each of the three groups develops a solidarity and places a distance between itself and the other groups. The squares tend to walk tall and smile, whereas the triangles tend to slouch. Even the controller seems to prefer the company of the squares to that of the other groups.

More trading sessions occur and then the controller says that, as the squares have been doing so well in their trading, they are to be rewarded for their skills by being allowed to change the trading rules if they wish. This is a touch of genius by Garry Shirts. The idea that those who do well in the game can then change the rules half-way through is mind-boggling.

What happens next is unpredictable. Usually the squares consult with each other in their group and decide to alter the rules in their favour with the aim of preserving or increasing their wealth. In the meantime, the circles and triangles are discussing the extraordinary turn of events in their groups and are becoming more and more indignant.

After this it rarely takes more than one or two brief trading sessions for the dissidents to mount a campaign of protest. At the right moment the controller steps in and, pointing to an angry, gesticulating member of the under-privileged groups, cries 'There is your revolution!' Or words to that effect. End of simulation. Everyone now realizes there was a hidden agenda. Many realize that the emotions and attitudes they have been feeling have been contrary to their normally stated beliefs. The squares realize that they have not only been using power, but also enjoying it. The triangles realize what it is like to be on the wrong end of injustice and inequality. The circles, who may have supported the class structure in the hope of bettering themselves and thinking that at least they were better off than the triangles, may be feeling rather guilty. And guilt often brings counter-attack — 'Is STARPOWER supposed to be an attack on our competitive society, on capitalism, on our democratic way of life?' etc. (It would be interesting to see STARPOWER used in a communist country.) Nor are the personal bitter episodes soon forgotten, even though they may be forgiven. Even if, at the crucial moment, the squares become altruistic and decide to alter the rules in favour of more equality of wealth, this in itself can result in a bitter and hostile reaction from the other two groups. 'Why should you be able to give us things — clear off!'

Sometimes the participants learn that violence is not the prerogative of children or non-intellectuals, and simulation literature contains some extraordinary descriptions given by the controllers or observers at STARPOWER sessions. Coleman (1977) gives an account of what happened at one session at a teacher training college in Cambridge:

> On one occasion a group of leftish liberal studies lecturers announced 'The name of the game is GRAB', and very shortly afterwards I was knocked to the floor and a pack of bonus cards torn from my hand. This was a pity — a meeting intended to show the hidden violence of our established society showed instead only the boorishness of some of its opponents.

STARPOWER is simple to introduce, highly ingenious, a remarkable opportunity for self-revelation, and as explosive as gelignite. It needs careful handling. The controller must be able to deal with the intense inter-personal hostilities which may well occur. As Garry Shirts says, STARPOWER should probably be used only by 'teachers who feel comfortable with vigorous reaction'.

DART AVIATION LTD

DART AVIATION LTD is another good example of a fully participatory simulation, and although very different from STARPOWER it shares the qualities of simplicity, ingenuity, and some tongue-in-cheek wit. Like STARPOWER, it was designed by an experienced author, Charles Townsend and, like TENEMENT, it is a British simulation. It is the third in a series by Townsend, entitled *Five Simple Business Games* — the other four being GORGEOUS GATEAUX LTD, FRESH OVEN PIES LTD, THE ISLAND GAME and THE REPUBLIC GAME. But although simple, it is not short and lasts about four hours involving roughly seven rounds of manufacture and trading.

It is intended for non-specialist pupils in the fifth and sixth forms of secondary schools. This, together with its simplicity, makes it rather unusual since most business simulations are somewhat complicated affairs, requiring several hours of pre-reading and designed for adults at management level.

The title is slightly misleading, as there is not just one company called DART AVIATION LTD but four, all starting with identical resources, operating under identical trading conditions and competing against each other.

Also very unusual is the combination of a business decision-making simulation with the manufacture of an actual product. The participants manufacture paper darts. This is a nice touch of humour, since paper dart making is the classic symptom of boredom. The darts must comply with specific quality control or they are rejected. They must be a specific length and width and fly at least four metres. Raw materials and tools are available at a fixed price. There are two sizes of paper and Sellotape is purchased by the metre. Scissors and rulers (the tools) may be hired at £1000 each for each manufacturing period they are required. Labour is also an expenditure. The controller uses a stop-watch or a watch with a second hand and the labour costs are £10 per worker per second.

DART AVIATION LTD has a nice balance between manufacturing and selling. Sales of aircraft depend on orders received, and orders received depend on what price each company puts on their aircraft and how much they spend on advertising. As in most other business simulations involving market decisions, the controller has arithmetical tables to work out the answers. Thus, if one company spends the maximum amount allowed on advertising and prices the aircraft at the minimum amount the rules allow, it will receive more orders than a rival company which spends less money on advertising and sets a higher price per plane.

If, however, a company receives more orders than it can fulfil, these surplus orders cannot be held over to the next trading period — they are lost. So a firm can sell no more than the number of planes it manufactured. If a firm manufactures a surplus of planes, these can be held over in stock until the next round.

From this brief outline — there is a lot more to it than this — it can be seen how a simple idea can generate a great many tasks which are common to most businesses: accounting, deciding how much raw material to buy, deciding on advertising and price, productivity, product design, training, etc.

Depending on the size of each team, it is recommended that individuals should allocate specific jobs among themselves — managing director, sales and marketing manager, accountant, worker, and so on. Since the motivation is high — the companies want to succeed — this division of labour within teams can enhance the participation. If the bright pupil finishes his task first, he has a strong incentive to help the others, and is unlikely to sit back and look bored. Although one role is 'worker', that does not mean that only one participant in a company can manufacture the aircraft. If two or more workers seem the best solution, then this can be done.

Similarly, if one team finishes its decision-making well ahead of the others in each round, it is unlikely to stop dead, since there are still problems to solve, methods of accounting to be explored, and future policy to be discussed.

Like STARPOWER, but unlike TENEMENT, DART AVIATION LTD has balance and fits neatly together with all the parts having both opportunity and incentive to work. The danger with DART AVIATION LTD, like other business simulations which have a series of trading sessions, is that the controller may let it run on for too long. There usually comes a time in such a simulation when the teams have solved their problems and worked out their best strategies, and things tend to become routine and repetitious. But this is a question for the controller to decide; it cannot be laid down in advance as the abilities and interests of different groups can vary considerably.

SPACE CRASH

I designed this simulation on the same lines as SURVIVAL, the first of my *Nine Graded Simulations* (Jones, 1974a), for communication skills. Whereas TENEMENT, STARPOWER and DART AVIATION LTD are about poverty or wealth in one form or another, SPACE CRASH concerns life or death.

The notes for participants is the first document to be handed out
and reads as follows:

NOTES FOR PARTICIPANTS

SPACE CRASH

What it's about

You are survivors of a space crash and it is your job to stay
alive.

Five profiles — Andro, Betelg, Cassi, Draco and Erid — explain
the situation. Read them carefully. They are important.

You have a map square made by Betelg which shows the area
in which your spaceship has crashed. Around the edges of the
map square is a description of what you can see one day's
walk away.

When you have decided which way to go, tell the controller
the number of the map square you want. Each new map square
shows you where you are and what you can see from that square.

You have a diary in which to keep a record of your day-to-day
movements from square to square. There are no diagonal
movements and you must all stick together.

Advice

SPACE CRASH may look like a game of chance but it is not.
If you treat it like a game, you are not likely to live very long.

Before you start deciding which way to go, try to make sure that
everyone has told you everything they know about the planet.

Don't waste time trying to think of ways to 'break the rules'.
For example, you have nothing to carry water in — and that's that.

It is a good idea to have a small coin or marker which you can
move from square to square as you travel across the planet. It helps
to show you exactly where you are and it helps you to keep an
accurate diary of your movements.

The profiles (or role cards) are designed to help the participants enter their space suits, as it were, and to encourage them to start talking. Here is the first role card:

Profile No 1: Andro

My name is Andro — space officer.

Our spaceship has crashed on Dy. We are in real trouble. Our radio broke down a week ago and cannot be repaired. No one will come to look for us here. Our captain is dead. Our food and water were destroyed in the crash.

Space officer Betelg has a compass and has made a map square showing where we have crashed. We must move, and we must stick together.

Dy is known as the planet of death. Dyans are friendly, and they would show us the way to go to reach a radio station which would have food and water and would rescue us. But Dyans do not drink water. They eat a kind of dry grass and they never move away from grassy areas.

Water is very important, but we have nothing to carry it in.

I will tell the others what I know. And they must tell everything they know about Dy, or we will surely die.

The controller gives the map square to Betelg. It looks like this:

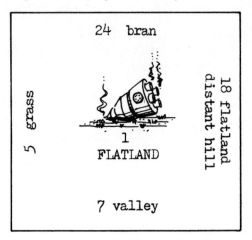

This is explained in Betelg's role card:

Profile No 2: Betelg

My name is Betelg — space officer.

I have made a map square showing where we have crashed — we are in flatland. I have called it Map Square No 1.

I have a compass. We can go north or south or east or west. But we cannot go diagonally on Dy — we would walk in circles.

We can see what is one day's walk away. There are bran sticks to the north, and we can eat bran. There is a valley to the south, and all valleys on Dy have water. There may be Dyans in the grassy area to the west. In the east is flatland, and across the flatland is a distant hill. It would take us two days to reach the hill. I have given a number to each square which is one day's walk away — any number will do — and I will make more map squares depending on which way we decide to go, and what we can see when we get there.

We must tell each other all we know.

The other three role cards are like the first two. Cassi's role card says they must have water every three days — there would be no fourth day without water, they would be dead. And the profile says: 'Someone once told me that sand on Dy is dangerous. I forget why. Perhaps the story is untrue.'

Draco's profile says they can eat bran which breaks off easily, does not go bad quickly, and can be carried around and lasts for months. They do not need bran straightaway but they must have bran within the first 15 days. And it adds 'How shall we decide which way to go — elect a leader, or take a vote, or what?'

Erid has 'made' a 20-day diary to record the day number, the map square number, and the sort of land they are on. Erid's profile explains that it is a 20-day diary because if they do not find a radio station in 20 days they will die, even if they do have water and bran.

From the controller's point of view, SPACE CRASH is incredibly simple. All the controller need do is hand out map squares on request. And even this is not necessary as the pack of map squares could be placed upside down on the table of each group with the numbers of the squares on the back of each card.

The main problem for the controller comes when death occurs which it does almost inevitably (about one group of adults in 100 survive). The controller usually says something like 'Pretend you did not go to that square. Go back one day, and you are still alive.' The controller retrieves the map square which led to death.

41

SPACE CRASH takes far less time than TENEMENT, STARPOWER or DART AVIATION LTD, but it provokes a high level of participation and a great deal of argument. The jigsaw nature of the information in the role cards starts the participants talking (and thinking), and the high stakes of life or death can induce even the shyest and most withdrawn student to speak out.

It is up to the participants to sink or swim or more usually to die of thirst. Death demonstrates to the participants that the controller will not intervene to help them in their decision-making, and that responsible failure is better than irresponsible success.

SPACE CRASH is the sort of simulation that often appeals to teachers of English as a foreign language, where a major problem is to encourage the students to talk. During the action the language used is that of opinions, hypotheses and conditionals — 'If we go east and there is no water, we will be dead.' And afterwards: 'What we should have done was . . .'

It can be seen that the four simulations, although very different from each other, treat the participants as responsible adults. All pose problems and all involve communication skills of one sort or another. In each case language is important. And it is the language of learning, not the language of responding with answers to show that something has been learned.

From the point of view of the next examination, however, it might be difficult to justify using any of these four simulations. 'What's the point of SPACE CRASH — my pupils are never going to crash on a hostile planet?' 'There aren't going to be any questions in the exam about making paper darts' 'If you think a lesson in violence is suitable for our sort of students . . .' 'He's not going to get a job as an old-age pensioner in a slum and the factual details of agency benefits are now out of date anyway.'

These questions relate to the objectives of education and are dealt with in the next chapter.

But the objectives of the simulation designer are not always apparent from examining the finished product. Authors rarely explain why they have put some things in, and it is even more unusual for them to explain why they have left other things out.

The wastepaper basket

Knowing what is left out can be of practical value to the teacher. Inexperienced adaptation can result in a teacher inadvertently re-introducing those items which the author has discarded because they did not work in practice, or because they unbalanced the simulation, diminished participant responsibility or caused unnecessary complications.

Another advantage of asking an author for an explanation is that this will make the useful point that it is often complicated to design something simple. It is a temptation to examine a simulation and say, 'Ah yes, that's straightforward and simple, so it must be quite easy to do.' Simulations are not easy to design and they are not easy to adapt — not, that is, without a good deal of practical experience.

Perhaps the most useful and revealing item in an author's workshop is the wastepaper basket. It will probably contain four or five times more material than the simulation itself. Many of the discarded bits of paper may be because the author failed to provide adequate precautions against the abdication of participant responsibilities. Participants, like everyone else, are prone to say 'It wasn't my fault.' If something goes wrong, they take full advantage of any opportunity to lay the blame at another door.

Consider the problem of helping an individual participant into an individual role. Here are two versions of a role card:

Version 1:

> **Role card: managing director of Blogsville Ropes Ltd**
>
> You are the managing director of a firm in Blogsville which makes high quality ropes. These ropes are used for a variety of purposes — for yachting, mountaineering, etc. You are a tough boss and have no time for troublemakers. In the coming negotiations with the trade union representatives in your firm, you will tell them forcibly that the company's present financial position is serious.

Version 2:

> **Role card: managing director of Blogsville Ropes Ltd**
>
> As managing director I shall be representing the company in the coming negotiations with our trade union representatives. Because our ropes are of high quality — used for yachting and mountaineering, etc — we have different problems from those of a manufacturer of a bigger range of ropes. Since the company's present financial position is serious, our negotiations will probably concentrate on this.

Assume that the role cards are handed out after the briefing and when the participants know the mechanics of the simulation and start to enter individual roles. In this situation 'I' is better than 'You'. It is realistic to suppose there is no need to tell the managing director of Blogsville Ropes that the product is of high quality and is used for yachting and mountaineering. Yet it is necessary to pass this information on to the participant. If this is not done in the briefing, the role card should convey the information indirectly if it is to avoid placing an unnecessary barrier between the participant and the role. Version 2 is an example of how this might be done.

Version 1 has a far more serious defect — it tries to imprint a personality as well as a function. Managing directors have the job of managing in the best interests of their companies. In a simulation this should be the aim of the person in the role of managing director. If Version 1 is used in a simulation, the following dialogue may occur in the de-briefing:

> *Controller:* Why did you, as managing director, break off the negotiations immediately the trade unions had put forward their demands?
> *Managing director:* In order to show them that I was a tough boss.
> *Controller:* Didn't you think it might be beneficial for your company to keep talking?
> *Managing director:* Oh yes. And if I had been the real managing director of Blogsville Ropes I would have done so.
> *Controller:* So what was the point of trying to demonstrate that you were a tough boss? Was that a responsible attitude?
> *Managing director:* No, it was quite irresponsible, but I was playing the part of a tough boss who had no time for troublemakers, and I thought that was what I was supposed to do. I wasn't being asked to behave in the best interests of the company, I was asked to be tough and uncompromising.
> *Controller:* Do you think the simulation tried to simulate reality?
> *Managing director:* Not really. It was a sort of informal drama. I thought it was intended to show what happened when the boss was a pig-headed power-mad nitwit.

This is the sort of conversation that an experienced simulation designer seeks to avoid and Version 2 is an attempt to block the loophole of evaded responsibility.

Blogsville Ropes is an imaginary example, but this sort of thing occurs frequently, creating an informal drama or pseudo-simulation rather than a genuine simulation.

Another reason for a full wastepaper basket is the problem of matching the simulation to the target participants. This is particularly difficult at secondary school level. There is a constant danger that the materials will be too sophisticated; on the other hand, there is a danger of over-simplification. If something is supposed to be a business letter, it should contain some business jargon and words of more than one syllable. The seemingly low level of ability of some secondary school pupils in tackling their first simulation is due to the fact that it is something completely new to them. They are not very good at tackling the unfamiliar — and this is true of most people. The first simulation is the most difficult. Afterwards it is easier, and the pupils become used to taking decisions and thinking for themselves, organizing themselves and working together for an objective.

One way out of the problem is not to over-simplify the materials but to increase the motivation. This is not easy and results in a lot of discarded work by the author. Here is one example — necessarily from the author's own experience — and it deals with a tailor-made simulation with specified objectives. This is a constraint upon the author, since it is quite common for an author to produce the simulation first and write the objectives afterwards.

In this case the Post Office wished to provide a simulation with the objective of encouraging young people to use postcodes and demonstrate their relevance to business. The primary target was 14 to 16 year olds who were following a commercial studies course, or an office practice course. The Post Office wanted the simulation to be simple and to have a 'right' answer rather than be open-ended. The most likely subject area would be the transport of goods by a company, provided it did not conflict with the sort of articles the Post Office delivers. The idea was to base the simulation on a manufacturer of confectionery, the materials would be postcode maps, a fair number of delivery notes giving addresses and orders for cakes and buns, and a number of transport schedules to be filled in by the participants, showing which of the half-dozen delivery vans went to which addresses in which order. It was thus a two-stage operation — locate the addresses on the map (or a facsimile for each pupil or each group) and then plan the delivery schedules. An additional idea was that if there were order forms as well as despatch notes, the pupils would be encouraged to file the documents which could be done on a postcode basis, this being an example of geographical filing.

These were the stated objectives, the recommended subject area, and an example of how it might be done. What, I was asked, did I think of it? Would it work? Could I design it?

My first thought was that I liked the briefing. The objective was clear, the target users were specified and, best of all, an example was given of how it might actually operate. My own preference is for open-ended simulations, but in this case it seemed appropriate that there should be a solution which could be easily verified by the teacher and appreciated by the pupils. This would, it seemed to me, mean that the solution to the schedule problem would involve the routes which used the least petrol — the shortest routes covering the delivery points but starting and finishing at the van depot.

Consequently, it seemed that the suggested example would work. It would be simple, it would probably achieve its aim, and the solution could be verified objectively. It would require a good deal of work to produce all the necessary materials but it was a practical proposition.

Having pursued this line of thought, it was necessary to start again. It seemed to me that there were really two major objectives:

1. to encourage postcode use, etc, and
2. to produce a fully participative simulation.

Full participation requires two separate ingredients:

1. the subject matter should be relevant, interesting, intriguing, or exciting, or emotional to the participants, and
2. the design should facilitate participation from all levels of ability in a group, avoiding excluding the slower learners who find the decision-making is taken over by the bright pupils; if there are groups of unequal ability, provision should be made for the bright group which finishes the job in half the time.

Looked at in this light, the example given seemed to suffer from certain defects. First, the introductory part of the simulation, the locating of the addresses on the map, ran the danger of becoming repetitious and boring, particularly since the motivation — the delivery of cakes — seemed rather lacking in inspiration. Second, and this was more of a problem, the solution of the route might well be found instantaneously by any pupil who was good at solving visual puzzles. Third, there was the problem of filing the addresses by postcode. This task could be given to whichever group finished first, or it could be incorporated into the proceedings earlier on. In either case it appeared to lack purpose — why bother to file them one way rather than another?

So I had a hard think about the sort of product that might replace the cakes, but which would not normally be delivered by the postal services. Images of various goods started to appear — refrigerators, furniture, bunches of flowers, hospital patients, and newspapers.

The next step was to take an example, dress it up, and see what it looked like. Newspapers. Weekly, daily, morning, evening, sports edition? Sports edition. Let's try it. Tentative scenario:

The participants are employees of a newspaper company, the Trumpet, a weekly paper in Bigtown which publishes on Tuesdays. The paper has decided to produce a Saturday afternoon sports edition. If that is successful it may expand and publish an evening paper every weekday. This would rival the other newspaper company in Bigtown, the Bugle, which already publishes an evening paper and a Saturday sports edition, and has its publishing offices across the street from the Trumpet.

It is vital that the Trumpet should deliver its sports edition to newsagents at the same time as the Bugle's sports edition, or preferably earlier, since newspapers are delivered on a sale or return basis. The first paper to arrive is likely to sell more, have fewer returned, and have bigger orders for future editions.

The Trumpet, like the Bugle, uses four of its own vans to deliver papers — north van, south van, east van, and west van. It is Tuesday afternoon, a slack time for the Trumpet, and the managing director has offered cup tie tickets and days off to any group of employees who can plan a good delivery route for each van. The employees are divided up into groups of four and each group is given maps and copies of delivery notes for the sports edition.

The employees are also told that the sales manager of the Trumpet thinks it possible that some areas around Bigtown do not have enough newsagents and, if this is so, something might be done about it. So he would be grateful if the employees would file the delivery notes in any order they think might help him examine the problem.

The scenario is sketchy, and would need polishing up considerably, even if it was accepted as suitable for development. And there are also the mechanics. The newsagents could be located in a way which allowed two solutions: a solution which uses the least petrol, and the 'nearest first, furthest last' solution which is the speediest delivery since the van is not delivering on the return journey. As one group is in charge of four vans, each participant can be responsible for one van, and when a bright participant has finished he or she can help the others. If a bright group finishes well before the others, the teacher can say 'It's now Saturday, but south van has broken down and only three vans are available.'

This somewhat lengthy and personal example has been given in detail in order to show detail. It is not intended to imply that newspapers are better simulation material than cakes — it depends on the circumstances. Moreover, the example concerns only the initial planning stages. Any of the details given could be rejected or altered. In fact, it is often necessary to alter the original concept because of problems which occur in the later stages of working out the materials themselves. In any case, what may appear to be a good idea may not work well in the classroom, and pre-publication testing is vital.

It is certainly the case that TENEMENT, STARPOWER, DART AVIATION LTD and SPACE CRASH were all thoroughly tested and the final package was probably the result of considerable alteration and adjustment. Unfortunately, this cannot be said for all simulations, some of which are published without ever seeing the inside of a classroom.

This chapter on design was not only intended to show how the bits and pieces of a simulation are made, but also to demonstrate that basically the authors and the teachers have the same aims and can work together in partnership, resulting in useful and smooth-running simulations.

Chapter 3

Choosing simulations

Problems

One of the main problems in choosing a simulation is that neither reading the instructions nor inspecting the materials can reveal the flow of interactions which may occur in practice. Inspecting a simulation package is not like reading a book. The best part of the simulation – the action – is left out. It is not available for inspection within the materials.

Not only is the interaction absent, but some of the best simulations contain clues, suggestions and options which are deliberately buried at various levels, and are not at all obvious to someone who merely reads the documents. Bits and pieces are put in to help the action; other bits are there to prevent it becoming one-sided or irresponsible. These finer points of design are rarely mentioned in the controller's notes.

The authors of simulations, like the manufacturers of motor cars, expect the customer to get in and drive – even for a trial run.

It must be stressed that in choosing a simulation there is no adequate substitute for taking part in it oneself. The controller should make every effort to participate – even partial participation is valuable. Corners can be cut, a role or two might be dropped, the action can be shortened, but even 15 minutes of physical participation is worth a month of reading the materials.

If it really is impossible to do this, observing the simulation in action is the next best thing. It will be an outsider's view, which is not at all the same as the insider's, but at least it involves action and the teacher can observe what the controller does.

Another possibility is to read the materials and take imaginary action in various roles, interacting with oneself, just as a person can play a game of chess without an opponent. But this is unsatisfactory and requires some experience of simulations. Not to be recommended is simply reading the materials as materials or, even worse, selecting the simulation from a list of titles.

It may be, of course, that the teacher has already heard good reports about a particular simulation, or has noticed that it is written by an author who has written other simulations which are familiar to the teacher. In this case, selection of such a simulation will be similar to buying a novel because it was recommended or because the reader liked other novels by that author.

Facilities checklist

Before shopping for a simulation it is useful and time-saving to make a mental inventory of the conditions and facilities which will be available. The greater the facilities, the greater will be the area of suitable choice.

MONEY

Financial considerations usually limit the number of simulations that can be bought and the price that can be paid. But within these limits the educational value of the simulation is not necessarily reflected in the weight of the documents or the gloss on the package. Like shopping for anything else, the teacher is looking for value for money, and this is a matter for the individual judgement of the teacher.

Generally speaking, good ideas are worth spending money on, even if at first sight the materials themselves seem somewhat scanty. If the ideas are good, the interaction is likely to be good, with the participants supplying by their behaviour and talk what at first sight appeared to be lacking in the documents.

Expensive simulations are not always good value, nor are cheap simulations necessarily nasty. On the other hand, if a good simulation is expensive, it is still probably better value than buying two less effective simulations at half the price.

However, the teacher should be told (or should find out) whether the materials are durable or whether they will be used up by form-filling, etc.

NUMBERS

Simulation instructions usually state the number of students for which they are intended. This is often a flexible figure. Also, the teacher will find in practice that most simulations can be adapted fairly easily to fit the specific number of students involved, by one participant taking on two roles, or two people taking on one role, or even occasionally omitting a role. Sometimes there are no individual roles, as in STARPOWER, and the size of the groups can vary considerably.

Sometimes a simulation can be run with two, three, or even four groups operating simultaneously, either as separate units as in SPACE CRASH, or in competition as in DART AVIATION LTD.

It is also possible on occasions to link simultaneous simulations. For example, there could be several appointments boards, but instead of each having separate candidates, the candidates could move from one interview to another. Opportunities for such flexibility will probably be mentioned in the controller's notes.

TIME

The materials should give an idea of the average amount of time needed and whether this includes the briefing and de-briefing.

As with numbers, time is fairly flexible in most simulations. If the time required appears to be too long in an otherwise suitable simulation, it is often possible to reduce the briefing time by a different form of presentation, or reduce the overall time by greater efficiency in the general running of the simulation. Similarly, a short simulation can be lengthened by various means, including a more thorough preparation, and an elaboration of some of the formalities and procedures within the simulation.

It may be a mistake to assume that the less able pupils need more time than brighter pupils to deal with a simulation. They may take more time to read and understand what they are supposed to do, but in the action part the brighter pupils spot more opportunities and will delve, prod, argue and discuss.

Time has a value, but there is a danger, as with money, in assuming that the less spent the better. Some authors argue the opposite. Elgood (1976) writes:

> Time is not at all a bad measure of value, and if certain items of knowledge are thought to be worth a major allocation of time, then the message of their importance tends to get over more strongly . . . It is too seldom admitted there is a positive correlation between the thoroughness with which people learn a thing and the effort they expend in doing it. Intellect is only one of the characteristics that is involved in the learning process. There must also be an emotional evaluation of the importance of the knowledge and this tends to relate to the emphasis placed upon it by the circumstances in which it is offered.

ABILITY

The main difficulty in assessing whether the materials are on the right level for the students' abilities is that these abilities cannot be assessed with any certainty without the students taking part in several simulations first. So if the teacher knows the students' abilities in simulations, all well and good. But if the teacher's views on student abilities are based on what happens in teacher-student orientated learning, the assessment may be inaccurate. Teachers who have had little or no experience of using simulations tend to underestimate the ability of their students to cope with the materials — not so much in the first simulation where there may be teething troubles, but in the subsequent ones. Even during the course of the first simulation it is often possible for the controller to notice the gradual (and sometimes not so gradual) increase in skill and confidence of the participants in handling the situation in which they find themselves.

The problem becomes somewhat easier if the simulations are graded. This technique is becoming increasingly popular, with simulations linked together like steps, the early simulations being relatively easy and the others progressively more difficult. Sometimes simulations are graded within themselves, with preliminary exercises, trial runs, or simplified versions of the main simulation activity.

Grading achieves two aims. It provides a greater flexibility for the teacher who is trying to match materials with ability, and it trains and educates the participants in the procedures and behaviour and skills required in the simulations. This means that there is a relatively greater return on the participants' time and effort than with 'one-off' simulations. Thus, DART AVIATION LTD is easier to get into if the participants have already taken part in the first two of Townsend's *Five Simple Business Games.* SPACE CRASH is easier if the participants have already died in SURVIVAL. In both cases, one simulation is similar in format to another and the participants know what they are supposed to do. The argument can also be applied to STARPOWER and Garry Shirts' other simulations. For example, BAFA BAFA has two hidden agendas, one for each of two groups representing different cultures, languages, attitudes and habits. Each group sends observers to the other group and, when the observers report back, they discuss and try to interpret the other group's culture.

MATERIALS

Some simulations require additional materials. These can range from paper and pencils to computers. They are usually listed in the controller's notes and are the sort of items which are readily available. However, there are often other materials which are not listed but which can be added to increase the realism or add to the effectiveness of different aspects of the simulation — a vase of flowers, an in-tray, a ten-minute break in the canteen for coffee, large sheets of card and felt-tipped pens. This point is dealt with more fully in Chapter 4.

ROOM SPACE

Room space is important in those simulations where several groups are operating simultaneously, particularly if it is desirable that they should meet in privacy (or secrecy). International affairs simulations and competitive business simulations are cases in point where an additional room or two can remove all temptations to observe or spy on what the other groups are doing.

However, if the simulation is right for the participants but only one room is available, then it is still possible to manipulate the geography of the furniture so as to lessen the chances of espionage.

Objectives

Just as authors sometimes design simulations, then find out what they have created and add the objectives on afterwards to fit the likely achievements, teachers might like to consider doing something similar.

This suggestion may sound like educational heresy, but it does have certain practical advantages for the teacher.

First, it means that the teacher's assessments and objectives are firmly based on observation — on seeing what actually happens. This is important in simulations, since what happens is often unexpected and may pass unobserved if the teacher is firmly concentrating on some predetermined objective.

Secondly, it is an attitude of mind which can broaden the spectrum of objectives. Instead of sticking to the obvious, it may encourage the teacher to look for other objectives as well and to evaluate all sorts of incidentals, which may seem peripheral but could be important, particularly to the participants involved.

Thirdly, it reduces the danger that the teacher will be conned by a high-sounding list of objectives contained in the publicity material for the simulation. Simulations have to work well to be effective, and objectives are not achievements. A good simulation with no objectives attached is preferable to a poor simulation with a thousand well-meaning aims.

Fourthly, the sort of additional objectives that are likely to arise from actual observation of what is going on in a simulation are probably those associated with education rather than training. Training objectives are easier to recognize and are more amenable to assessment. But educational objectives — the abilities and skills which make up so much of adult life — are not easy to recognize and may be impossible to quantify. But that does not mean they are not important. Confidence, for example, is tremendously important. So are organizational skills, the ability to use language and communicate effectively, and all sorts of personal traits and attitudes which help people to get on with one another.

In a simulation, important changes in a person's thinking and behaviour can occur and yet pass unnoticed because the teacher is concentrating on whether, for example, the participants are appreciating the problems involved in the use of land resources.

The argument is not that objectives are unimportant, rather the reverse. Objectives are so important they should be constantly appraised in the light of observation and experience. If a teacher finds that a particular simulation is consistently successful in achieving some unexpected yet desirable learning behaviour, this can be added to the list of objectives. If expected attainments do not occur, these can be dropped from the objectives. However, apart from choosing a suitable simulation and presenting it effectively, there is nothing the teacher can do to ensure that the objectives are achieved; that depends on the participants and whether the objectives are appropriate.

It can be useful, therefore, to examine some of the objectives which authors have listed as important, and see how these may influence a teacher's choice of simulations.

FACTS

Many situations, perhaps most simulations, are aimed solely at conveying facts and insights about a particular subject. Some of these are training simulations, designed to familiarize the participants with the subject and the procedures for dealing with it.

Some are tailor-made simulations for a specific company or organization and are intended to give the students a feeling of what it is like to be doing a specific job. These, however, tend to be guided exercises rather than genuine simulations, and in any case there is little question of choice involved as they are often so specialized as to allow no adequate substitute.

Leaving aside tailor-made simulations, the advantage of using simulations to convey the facts is that a more personal and committed form of learning may result. This supplements but does not replace the textbook. In addition, such factual simulations are easy to justify in the classroom and staffroom, as they are likely to be directly relevant to forthcoming examinations.

But of all the objectives which a teacher may have, the conveyance of facts about a specific subject is the most limiting as far as making a choice is concerned. If the teacher wishes to introduce a simulation about a specific event or historical development — the Wars of the Roses, for example, or the growth of railways in Africa — there may be few if any simulations to choose from. As there is now a rapidly growing number of simulations being published, however, it is worthwhile for teachers to delve into directories of simulations and publishers' lists to see what is currently available. As with any other form of learning materials, it is up to the teacher to assess the quality of the facts provided — their accuracy, their significance, their presentation — in the usual way. The teacher must exercise professional judgement and take into account the reputation of the author and publisher.

If the facts are contemporary, not historical, the teacher faces an additional problem — the facts may be (or may become) out of date. This is particularly the case in simulations like TENEMENT which deal with the specific and current social services, agencies, benefits, laws and regulations. Since such simulations convey an important feeling of immediacy and relevance in the areas of personal issues and general problems, students may well conclude that this is how things actually are in the details about pensions, job security and housing. The teacher can look at the publication date and try to update the materials or can explain that certain facts are no longer true; alternatively it can be presented as an historical simulation.

Choosing a simulation for its facts is not the same as choosing a textbook for its facts. In a textbook, the facts are there to be learnt; in a simulation, the facts are there to be used. If in a simulation the participants decide that some of the facts are irrelevant to what they are trying to do, they may dismiss these facts with no more than a quick glance. If the simulation has a huge scenario, with page after page of background facts which have little relevance to the action, not only may the facts go unlearned, but a feeling of annoyance may arise which has a negative effect on fact learning.

If, on the other hand, most of the facts presented in the simulation are potential weapons in a conflict of interests, they are likely to be scrutinized far more closely than would ever occur in reading a textbook. As well as learning relevant facts, the participant would be developing valuable skills of selection, analysis and presentation.

MODELS

In simulation literature the word model does not imply miniaturization, in which all the details are reproduced but on a small scale. When talking about simulations, the word model is used to cover the essential working elements of something — an economy, a political system, a society.

If the teacher's objective is a model, the choice is far wider than if the objective is to convey specific facts, because models do not have to be based on actual facts. Sometimes fiction and fantasy can provide a clearer picture of the way things work than an attempted simplification using actual organizations and specific examples. Fiction can be manipulated by the author to highlight essential aspects in a way which may be difficult to achieve with actual real-world components with a lot of historical and perhaps irrelevant clutter.

There is no right answer to the question of whether to choose simulations with actual names, organizations and places, or whether to choose simulations using fictitious elements. The familiar has the advantage of being easily identifiable; the unfamiliar has the advantage of giving the students an opportunity to take a fresh look at an old problem.

A model based on fiction may have to go into detail about procedures and background simulation 'facts' which might be unnecessary if actual organizations were used. On the other hand, a model based on actual names and places may run into the problem of which 'facts' are real, and can bias the outcome of the simulation in favour of those students who happen to know the most about that specific organization, country, or period.

DECISION-MAKING

Decision-making covers a wide variety of skills and disciplines. It is not the same as decision-giving; it is not just saying 'yes' or 'no'. Making decisions involves searching around for the most suitable decision, analysing the situation, and constructing hypotheses about what might follow certain decisions.

One of the problems of training is that people are trained to solve the same sort of problems and make the same sort of decisions again and again. This is why many courses in business management deliberately seek out unusual problems in order to broaden horizons and shake up established thought patterns.

So, in having decision-making as an objective in choosing simulations, the key question is the sort of decisions that are required. Are they specific, or are they general? Should the decisions be unfamiliar in order to give students practice in coping with the unfamiliar, or should they be restricted to previous decision-making experiences?

An advantage of unfamiliar decision-making is that it changes the existing classroom hierarchy which is usually dependent on factual knowledge. With new problems students have more equality of opportunity. For example, in STARPOWER, the squares are faced with a completely new problem — how to alter the rules. The situation has never arisen for them before. They cannot use previous knowledge. In SPACE CRASH the participants have never before been faced with a decision about whether to go north to the bran or south to the valley. Usually they would have maps or some knowledge of what is one day's walk away but not in this case.

In DART AVIATION LTD decisions are on familiar ground. Many participants will already have had experience of the problem of manufacturing paper darts. Decisions on prices, advertising expenditure and aircraft manufacture in the simulation are the same as they would be with a thousand other products. The decision-making is not a matter of life or death, but of gradual adjustments, trial and error, hypotheses, and analysis of results which can be thought of entirely in numbers. Ethics are not involved. The problem is related to maximizing profits, and this is true of probably most business management simulations based on a 'model' — an arithmetical model for determining the results of business decisions. If the participants have already had experience of this type of problem and this sort of decision-making, DART AVIATION LTD will help reinforce the learning. If, on the other hand, the participants are new to business simulations, the experience will be an extension of their personal experiences of buying and selling and advertising.

Again, there is no 'right' answer about whether to choose the familiar or the unfamiliar problem for decision-making. The familiar may be more related to training and the particular; the unfamiliar may be more related to education and the general. But circumstances alter cases and it is up to the teacher to select the most suitable objectives.

The same consideration applies to open and closed simulations. In arithmetical-based business simulations, the arithmetic is an objective criterion for measuring the success of the decision-making. The problem is to crack the code and this is done by analysing the result of round-by-round decisions.

In closed simulations there may be one right answer or several, but there is an objective criterion which enables students to say 'Oh, yes, I see, we got it right (wrong).' Open-ended simulations, on the other hand, may not only lack any objective criteria, but may also be deliberately designed to provide simulation 'facts' of equal weight to conflicting interests to balance arguments and make everything a matter of opinion. Behavioural and human relations simulations tend to be open-ended; so do conflict simulations in local government or international affairs. In open-ended simulations the decision-making is not a neat and tidy scientific affair. Thus, STARPOWER is unfamiliar and open, whereas SPACE CRASH is unfamiliar and closed. STARPOWER is unpredictable — almost anything can happen — and who is to say whether what happens is right or wrong? In SPACE CRASH the participants either live or die and almost always die.

However, in a simulation, unlike an examination, the result is rarely important. What matters is how the result is arrived at. Secondary school pupils tend to prefer closed simulations because they satisfy a 'What's the answer?' approach to learning. Whether this is a good enough reason for a secondary school teacher to select closed simulations is another matter. The teacher may take the view that one objective is to try to persuade the pupils out of the habit of asking 'What's the answer?' to anything and everything. In many areas of life 'What's the answer?' is not an appropriate question. Questions themselves can be more important than answers. Exploration can be preferable to arrival. Decision-making can concern opinion as well as fact.

COMMUNICATION SKILLS

Through effective communication, we deliver our ideas. It is useful to be able to communicate and even better to be able to communicate effectively. It is inefficient for brilliant ideas to be concealed by inarticulate mumbling.

One of the great advantages of simulations is that they are self-activating and provide scope for a far wider range of useful communications than normally occurs in education. Instead of the student communicating with the teacher, usually answering questions to demonstrate that something has been learned, in a simulation the student learns *by* communicating. In TENEMENT the participants talk in order to give or receive state benefits; in STARPOWER they talk in order to increase their wealth or change the rules; in DART AVIATION LTD they discuss business and manufacturing strategies; and in SPACE CRASH they talk about their problems in order to live.

There is a wide range of communication skills — diplomacy, arguing, interviewing, chairmanship, reporting, presenting a case, speaking in public. In many schools pupils go in at one end and come out at the other without any personal contact with many of these skills. Practice brings skills, and skills bring confidence; without practice there is no confidence. Many students hate the idea of speaking in public and would run a mile if asked to be a chairman or to present a case. Yet these skills have a wide transferability. The skill of diplomacy is of value to all people, not just diplomats. The ability and confidence to speak in public are of value to others who are not public speakers. Being able to speak in public does not mean that one has to do so; but it can be most satisfying to know that one could cope if the occasion arose. Obversely, lack of practice, lack of skills and lack of confidence can on occasion result in near mental paralysis if a person is afraid of being called on to speak.

If practice in communication skills is one of the objectives of the teacher in choosing simulations, it is probably a good idea to seek out simple, argumentative and emotional simulations as starters. Some simulations are specifically designed to encourage communication skills and all simulations involve communication. So it may be worthwhile looking at the difficult part of a simulation — the bit that is left out — to see what sort of communication is involved. If there is a series of simulations on the same subject, this could be a gradual way of giving communication practice. Alternatively it may be better to look for a wider variety of roles and situations to cover more skills.

Fortunately, the specific subject is not all that important since it is difficult to be clear and articulate in one subject without also being good at communicating other subjects too. Preparing one speech helps to prepare the next one, even if the subject is different. Diplomacy in international simulations may help diplomacy in family and personal relations.

LANGUAGE SKILLS

Language skills are allied to communication skills, and sometimes the two phrases are used interchangeably. In practice, language skills are usually taken to refer to the sort of skills taught by the English department or by teachers of English as a foreign language and these are not identical.

Yet, in both fields, the first problem may be to encourage the students to talk. One teacher in a north London secondary school observing his class take part in their first simulation, said with astonishment, 'I've never heard half of them speak before.'

It is not just a question of language, but a question of confidence. Simulations have the advantage of removing the teacher, who is sometimes an inhibitive figure, respected yet feared. To the shy student it may be better to keep quiet rather than say something wrong and be laughed at.

The British Council is in the forefront of advocates of simulations. Kerr (1977) says:

> They ensure that communication is purposeful (in contast to the inescapable artificiality of so many traditional exercises and drills); and, secondly, they require an integrative use of language in which communicating one's meaning takes proper precedence over the mere elements of language learning (grammar and pronunciation).

If language is high on the teacher's list of objectives, subject matter is probably a good deal lower down. It may be a prescription for failure to set out with the aim of matching the students' interests with the subject matter of the simulation. If the teacher has a group of Spanish engineers as students, it is courting disaster to place a blind order for a simulation that is about (a) Spain, (b) engineering, or even (c) Spanish engineering. The scenario may be bulky, the language content may be slight, and the interaction virtually non-existent. The main language skills may come afterwards when the participants complain that 'It is not like that' in their branch of engineering or in their part of Spain.

BEHAVIOUR

Quite a number of teachers introduce simulations for exclusively behavioural objectives and these can vary considerably.

One is the straightforward objective of improving the behaviour of the class. If there have been antagonisms, frustrations and tension among pupils, boredom or dissatisfaction with the course, or potential hostility between pupils and the teacher, a good simulation can work wonders. It removes boredom, redirects the activity, and extricates the teacher from a confrontation.

Other behavioural objectives can range from self-awareness to an examination of the hidden motives and attitudes in society. By their nature such behavioural designed simulations tend to be controversial on all sorts of levels. For example, Zuckerman (1973) notes that most racial attitude simulations ask the participants to get into the black experience. He says this encourages 'shallow depth responses along the line of "Lo, the poor black man, who is so mistreated by those *other* people."' Shirts (1970) says that simulations about the black community are generally written by people from the suburbs and are based on a series of unfounded clichés about what it is like to be black, which not only encourages stereotyping but creates an attitude of condescension towards blacks. Also, says Shirts, such simulations can give the students the impression that, having taken part in the simulation experience, they know what it is actually like to be discriminated against or what it is like to be black.

These criticisms are not altogether satisfactory. For example, if there is stereotyping, this suggests that it is not a simulation but an informal drama. If a role card for a member of the housing department says, 'Generally does not favour allocating houses to blacks in white areas', this is personality imprinting – a stereotyping which denies the participant the right to make up his own mind in the light of the situation. Similarly with the question of condescension, unless it is role-play in which the person is asked to behave in a condescending manner, condescension is no more predictable than sympathy or respect. In a genuine simulation, the participants can feel any way they like, providing they do their job as best they can.

Behavioural simulations, like textbooks for filmstrips or television programmes, may or may not be typical of specific problems. But no medium exists in isolation and any individual example in any medium is unlikely to convince a student worth his salt that he 'knows it'.

In addition to the 'big issues' there are plenty of what might be called 'local' behavioural simulations. Interview simulations are a case in point. The objective is to help the participants behave more effectively in interviews.

There are personal conflict behavioural simulations – conflicts between boss and employee, between parent and child, between teacher and pupil. Some of these are close to role-play exercises, but others are genuine simulations involving groups of people, documents and decision-making. The dividing line is not so much the number of participants or the documentation, but whether it is teacher-guided, whether it is personality-imprinted, or whether it is based on job function.

FRIENDSHIP

Friendships develop in simulations and there is no reason why they should not be added to a list of objectives. Indeed, some teachers use a simulation for no other purpose than to help students get to know each other at the beginning of the academic year. Not only does it help students get to know each other, it also helps the teacher to get to know the students in a way which is different from restrictive teacher-student orientated behaviour. With simulations it is quite common for episodes to occur which take the teacher by surprise. 'I had no idea that Mandy had it in her' is a common remark. Consequently, the teacher has learned something, and so probably has Mandy and the other students.

Friendships involve more than just friendly feelings; they imply understanding and communication, and working together effectively. All these can be very useful at the beginning of a course, and pay dividends later on in work which might be completely dissimilar from the subject matter of the simulation. Teachers may tend to undervalue friendships, thinking them personal, not educational, but they are very important in education just as they are in the world outside the classroom.

PREDICTION

This category of objective has to be mentioned since it is extremely valuable to certain specialists who wish to find out in advance how certain things are likely to work.

Foreign ministries may want to know 'What is likely to happen if we did so and so?' A simulation is the best method of finding out, short of actually doing so and so. A defence ministry may wish to know the most effective strategy for some newly developed weapon — but it cannot start a war to find out. There is no alternative but a simulation. Local authorities may wish to know if their resources for dealing with a disaster are likely to be effective. Again, it is a matter of simulating the situation. But prediction is not an objective that the practising teacher need consider. If it were worth considering, then almost certainly the teacher would already be using predictive simulations.

SELECTING

The bulk of the work of selection has been done once a teacher has assessed the resources available and has settled on the objectives.

All that remains is the practical step of examining and trying out potentially useful simulations.

The additional points to watch for have already been dealt with in the first two chapters — the need to look at the design of the simulation, to see if it is fully participative, if it is well balanced, if it is provocative, stimulating, interesting, emotional, involving, and so on.

The basic question is whether it will work, given, of course, the right sort of participants and an adequate introduction by the controller. As stressed earlier, selection does really require preliminary personal participation by the teacher in one form or another. Once that has been done satisfactorily, what follows is relatively plain sailing.

Chapter 4

Using simulations

Participation

If the teacher participated in a simulation at the choosing stage, all well and good. But if not — perhaps because the simulation had already been chosen and was awaiting the teacher's use — the teacher should arrange a participation session with a few friends or colleagues.

Some teachers protest at this advice, because they see it as a waste of time. They see the simulation in the same light as a book or film and think that all that needs to be done by way of preparation is to inspect it. If it is not possible to persuade or bully people into having a complete run-through, at least the teacher should make every effort to enrol a few people for participatory sampling, as suggested in the last chapter.

However, if the teacher is an expert controller and has used similar simulations before, obviously less preparation is needed because the teacher is more aware of what to look for in the materials, what to aim for and try to avoid in the simulation, and is more experienced in briefing and de-briefing.

In being controller, as in everything else, practice brings skill and skill brings confidence. The expert simulation user can inspect materials and say something like this: 'Now, I am participant X. I have read the materials and documents which relate to my role. Am I a member of a team or an individual or both? Do I have to take a decision immediately and, if so, what sort of decision and what are the possible consequences? Can I and should I wait until someone else has taken the initiative? What options are open? Is it clear what my responsibilities are and what my powers are? Do I know what I am supposed to do? Are there any other participants who are my natural allies or enemies? What, if any, are the physical things I must do or might do — such as filling in decision forms, writing requests for meetings, voting, etc?'

With something like an arithmetical model-based simulation the teacher should actually fill in a decision form and process it — perhaps with other group decisions — in order to arrive at the resultant number which comes out after the data have been fed through the arithmetical formula.

This should be done not once but twice, or as many times as may be necessary until the teacher is sure about the procedure. A decision-making form is designed to lead the participants through the 'rules' in a step-by-step fashion. But often things go wrong. The wrong information gets into the wrong box. Misunderstandings occur. Errors creep in. In this sort of simulation, where the teacher as controller will be responsible for processing the decisions, it is essential that the teacher should prove, in practice, that he knows how it works and also acquire some experience of the 'right' sort of decisions. If the teacher feels it is unnecessary to go through the chore of physically filling in the forms in the belief that he understands the instruction, then disaster moves from the category of 'possible' into 'probable' or 'certain'. The sort of thing which could happen is when a group has been filling in forms wrongly without knowing it and so receiving the resultant 'wrong' answers. By the time the students have acquired enough experience of the simulation to realize that something is wrong, it may be too late to unscramble the mess. Everyone will have to start again or abandon the project — and either eventuality will do nothing for the teacher's reputation. Reading the instructions and saying 'I don't need to try it, I understand it' are famous last words.

Controller

The most important role in a simulation is that of controller (arbitrator, umpire, instructor, etc). It is the controller who is solely responsible for the briefing, the de-briefing, and the mechanics of the action.

The first job of a novice controller is to shake off any inappropriate habits or thoughts derived from the experience of teaching or instructing. The main part of every simulation, the interaction, is not taught. By training and by habit teachers interrupt, guide, explain, give hints, smile, frown, and in many subtle ways (including silence) try to help students learn. But if a teacher tries to do this in a simulation, it stops being a simulation and becomes a pseudo-simulation or a guided exercise.

Even during the briefing and de-briefing it is useful if the relationship between controller and students is that of the relationship between professionals — respectful, polite, slightly distant but enthusiastic about the speciality — in this case the control of the mechanics of simulations.

Once into the action, the controller should aim to be invisible, to merge into the background or to assume the protective colouring of a plausible role such as usher, messenger boy, furniture remover, or friend of the editor or managing director. When it becomes necessary to speak to the participants, it is most valuable for the controller to adopt appropriate protocol. Instead of saying 'If any student . . .' or 'If any participant . . .', it is better to use the appropriate titles — 'If any Honourable Members . . .', 'If any executives . . .', 'If any councillors . . .', 'If any journalists . . .'

As mentioned in Chapter 1, the concepts of reality and power are vital, and these provide guidelines for the controller and for the participants. Reality for the participants is the interaction. It is theirs; they own it. Conversely, what happens (or is supposed to happen) outside the area of the simulation is the responsibility of the controller, or the control team if it is a big simulation. This means that if any participant orders an event to happen outside the room — calls for a day of prayer, calls for troop mobilization, calls for a strike, calls for a protest march, calls for the police — it is up to the controller to decide what happened and to inform the participants accordingly in whatever manner seems appropriate within the realism of the simulation.

Some simulations have their own built-in mechanism which makes this unnecessary. In management simulations, and within specified limits, certain orders are automatically carried out according to the mechanism. Providing the companies in DART AVIATION LTD fix their advertising costs within certain limits and in multiples of £1000, this advertising will occur. There is no chance of the advertising agency going bankrupt or the advertisements being declared illegal or ineffective; that is not part of the simulation. Nevertheless, should any group of participants wish to take decisions outside their area which are not covered by the 'rules', it is a matter for the controller alone to decide what happens.

The participants have power within the simulation interaction; the controller has power outside it. Power, authority, duty and responsibility are thus clearly defined and are clearly separated by the line between inside and outside.

Detachment

Within a good simulation teachers are often subjected to the strongest temptations to participate themselves. In the swirling action, the vigorous arguments, the emotional, imaginative and momentous events, the controller can sometimes be observed grinning with excitement and holding himself back from rushing in to participate. Extraordinary though it may seem, the teacher sometimes does actually sweep forward and assume some role or other. This is bad intervention. There can be good intervention if, for example, a key role is obviously vacant. Perhaps there should be someone in the role of Secretary-General of the United Nations for a few moments, or a policeman, or a king, or an usher. These are temporary interventions to assist the mechanics, rather than interventions for personal reasons. It is a general and a safe rule never to interfere, nor to give any signs of pleasure, displeasure, surprise, boredom, annoyance, appreciation or exasperation.

A poker-face is virtually obligatory. It is not easy but is worth cultivating since it deters students from looking at the controller for signs of acceptance or rejection. Poker-faces help to preserve participant power and responsibility. At first this may seem strange and unusual to the participants, but they soon see the value of it and take pride in 'owning' their simulation, which would be impossible if they had to keep looking over their shoulders at the controller.

This detachment by the controller could be explained in the briefing: 'My job is to remain poker-faced. There should be no point in your looking at me. Should you see me smiling or frowning, I am not doing my job properly. Try to take no notice of me at all. A simulation is not like a play; you are not going to be coached and guided and stage managed. We are not aiming for a perfect performance. You are on your own. You've got a job to do. Just do your best and ignore me.'

Observing

With some simulations the controller has no problems about observing what is going on without interfering in the action and without disrupting the participation. If the simulation is a public debate, the controller can sit poker-faced in a corner of the room and can see and hear just as much as any of the participants. But a problem arises when groups meet separately and in secret. Should the controller pull up a chair and say, 'Don't pay any attention to me' and listen to what is being said? With three or four groups, the controller could move about and enter the intimacy of the circle of confidential talk, and then move off and on to another group. Is this a good idea? Is it a good idea even if the controller resists all temptation to smile, frown, look appreciative, perplexed, or whatever?

In general, the answer must be no, it is not a good idea. If it is a straight choice between controller ignorance and controller interference, it is better to remain ignorant rather than risk distraction or interruption or subtle manipulation.

This is because what happens happens, and the interaction should be the sole responsibility of the participants, without interference. It can, of course, be recollected and discussed in detail at the de-briefing. The group can say, 'We had a lot of trouble deciding whether to do so and so, and Mr X said such and such and Miss Y replied . . .' The fact that the controller did not listen to the actual conversation is not important; what is important is that it took place.

This advice may be very difficult to take. Teachers like to know what is going on. If they know what is going on, they can use the information afterwards and give advice, hints and guidance in the de-briefing. It is all part of the teaching technique to observe and comment.

But a simulation is not teaching. It is learning and the learning will be placed in jeopardy by interference. The learning is not just the learning of facts; it is behavioural and concerns power and responsibility and should not be diminished. No teacher walks into a cabinet meeting or boardroom meeting and says, 'Don't mind me, I'm not here, just carry on normally.' If such a remark is to be made, it should be made during the briefing, not in the middle of the action.

There are, however, certain techniques for observing without interfering. One is to adopt a slow pattern of repetitious behaviour which brings the controller within earshot of what is going on. For example, instead of the controller sitting at the control desk while three or four groups have private discussions, he can walk very slowly and majestically around the room, with head raised, staring straight ahead and seemingly oblivious of what is going on in the discussions. Provided the controller does not pause, hesitate, or stop, the participants will soon get used to the perambulations and ignore them. Using this sort of technique it is possible for the controller to listen to quite a lot of what is going on without appearing to do so.

Role allocation

Teachers who are experienced in using informal drama and role-play exercises probably have their own favourite method of allocating roles. But a simulation is quite different from an exercise and an informal drama. Different considerations apply and what may be suitable for the one may be unsuitable for the other.

There are two kinds of roles — group roles and individual roles. These can be considered separately even though there are many simulations which contain both forms.

Group roles

It could be a group of businessmen, statesmen, tribesmen, or whatever, but the key question is which group the individual participants should belong to, rather than what their function will be as members of that group.

One question is whether students A, B and C, who always work together, should be allocated to the same group. Another question arises if groups have important distinguishing characteristics which evoke preferences or prejudices. For example, who should belong to the Bosses' Party and who should be members of the Workers' Party? Role allocation can be random or by individual preferences, or by teacher selection. Each has advantages and disadvantages.

Teacher selection has the advantage of putting the decision-making into the hands of the existing classroom authority — the teacher — who should be best placed to make the selection on whatever criteria may appear appropriate. The teacher can keep together a group of friends or can split them up. Deliberate selection allows the teacher to restore the balance of arguments in a simulation by placing the more able students in that group which is the least popular among the students, eg the chemical company management rather than the anti-pollution protesters.

There are potential hazards in teacher selection of group roles. One is that it may arouse resentment among the students; they may feel manipulated or discriminated against. If so, this would be contrary to the basic principle of simulations which is to give as much power and responsibility to the participants as possible.

Another problem is that if the teacher and students are new to simulations, the criteria of student ability may be inappropriate, since it would be based on normal classroom behaviour and results.

Surprising things can happen in simulations and the usual hierarchy can be rudely shaken.

Probably the easiest method of allocation to groups is to keep together those students who usually work or play together. This still leaves the problem of bosses or workers, landlords or peasants, hunters or farmers, traditionalists or revolutionaries, but it does mean that during the course of the simulation there is a minimum of friction within individual groups. Naturally, the strength of this arrangement depends on the strength of the group feelings in the class.

The random method of group allocation offers a solution to several problems simultaneously. First, it is fair and can be seen to be fair. Whether student A is a landlord or a peasant depends on what it says on the piece of paper he picks out of the hat. This avoids any accusations of teacher favouritism or teacher manipulation. The groups should average out in simulation ability and the larger the number of participants the closer each group's simulation ability is likely to be.

The main disadvantage of the random method is the resentment felt by friends who find themselves separated. You could explain, however, that this is one of the hazards of life, and that it is a good idea to try to co-operate with people who may not be friends and may sometimes be enemies. This disadvantage is likely to diminish as students widen their circle of acquaintances and friends, which could turn a short-term disadvantage into a long-term gain.

It is worth making the point that random selection really should be seen to be random and not an announcement, 'I'm randomly allocating student A to the king's party and student B to the slaves' compound.'

Some simulations can be re-run with a change of group roles in which case role favouritism is no longer an issue, and this leaves only the problem of the composition of the teams.

Individual roles

An individual role means that an individual has some particular function, responsibility or knowledge which is different from the other participants.

In TENEMENT, although there is a group of seven tenants, they are all individuals with individual role cards, individual problems, and individual circumstances.

In DART AVIATION LTD, there are individual roles but no individual role cards. The roles are allocated within each team, by each team. Therefore no individual possesses any knowledge not also available to the whole group. In this sense the roles are simply an allocation and division of responsibilities and functions.

In SPACE CRASH, the names of the space crew and the individual roles are mainly for sharing out information (rather than being embodied in common 'rules'), and for encouraging communication of ideas and giving an atmosphere of impending doom.

In STARPOWER, there may appear to be no individual roles, only group ones, but this is not quite the case. Those individuals who are fortunate, or unfortunate, to receive more wealth, or less wealth, than most of their group are from the start predestined for promotion or demotion.

For the teacher, a key problem is what to do if a simulation has a role or roles implying specific skills, such as a technical expert. An important consideration is whether the scientific data are available to all the participants, perhaps in a 'library', or are restricted to the profile or documents belonging to the expert. It may also be that the public documents contain only general scientific evidence, while the individual profile and private documents give additional evidence in detail.

Other considerations are whether the scientific evidence is disputed. Are there two or more 'expert' views, or does the simulation provide only one undisputed view? How important is the scientific evidence in a given simulation? Are the overriding factors public opinion, financial priorities or statistics about acceptable or unacceptable levels of pollution?

Answers to these questions will help in deciding how to allocate the role of expert. If individual roles are allocated at random and student X or Y or Z gets the role of expert, does that unbalance the simulation or render it less effective? Or are there safeguards? Can other participants unearth the information and question the testimony of the expert? Does it matter much anyway since the educational aim is likely to be practice, not perfection?

Cases should determine decisions. But it is probably a good idea, in those simulations where the role of expert is of some prominence, to allocate the role at random but to allocate two participants to the role. They will back each other up and it is not unrealistic for an expert to have a colleague or assistant. This technique can also be used in simulations which require the role of chairman. It can be the role of chairman of the national assembly or the public inquiry, or a sub-committee of the local boys' club. Even if all the participants are themselves top-level adult experts, it might still be a good idea to consider whether it would be desirable to allocate a second person to a role — perhaps deputy speaker, deputy leader, deputy chairman, deputy prime minister.

The arguments in favour of randomness in individual role allocation are similar to those already given in the allocation of groups. The technique is fair and is seen to be fair. People can develop by having responsibility, and the aim is to give opportunities not to engineer perfection. Random allocation enhances the principle of participant power and strengthens the role of controller, since it emphasizes controller impartiality. As with group allocations, the main disadvantage of randomness in individual allocations is likely to be in disrupting the pattern of friendships or working acquaintances. But this is likely to be temporary and may well have longer-term benefits. The most difficult simulation for random role selection is always likely to be the first simulation; afterwards it becomes much easier.

An alternative to both random and teacher selection is to allow the students to volunteer for specific roles. In practice, this is not usually a good idea because the simulation is likely to start with disputes or feelings of unfairness. The shy student may feel resentful that his own inhibitions prevented him from getting in first with his application for a specific job, and he may be landed with a job at the lower end of his preferences. The extroverts and top dogs will have the satisfaction of getting the best roles — or more probably the easiest roles — and this may also be accompanied by feelings of defensive guilt. These emotions can surface during the simulation or remain like hidden rocks which can damage or sink the simulation without the controller being aware of them.

Presentation of materials

What should be a simple matter of presenting the right materials at the right time seems to cause far more trouble than it should. Possibly one explanation is that the teacher has not participated in the simulation personally and yet is under the impression that he 'knows it' because he read the instructions last week or ran the same simulation last term.

The need for care in presenting the materials is not simply because it is more efficient to do it properly. Another reason is that in a simulation the penalties for mistakes are likely to be greater than in traditional teacher-student situations. If, in normal classroom discussion, the teacher has to fumble around to find the right bits of paper, this may be rather annoying or amusing to the students, and they probably do not object to the delay. But in a simulation, which tends to develop a considerable dynamic quality of its own with a great deal of involvement, any delay caused by controller inefficiency may well result in considerable exasperation from the participants. Furthermore, the mechanical breakdown may jerk the participants out of their roles, and it may take them some time to re-establish the flow and realism of the simulation.

The actual loss of a document can unbalance the simulation to such an extent as to make it virtually unworkable. In ordinary classroom teaching, the loss of one fact-sheet out of ten may not be disastrous, but in a simulation, where the pieces fit together and are often interdependent, the loss of one part can easily nullify the whole. So the teacher must allow sufficient time to see that all the required documents are available and that they are arranged in the right order for presentation.

In preparing and arranging the documents, the teacher should consider not only the action part of the simulation, but also the briefing. Are any additional documents desirable in order to clarify the basics of the simulation? For example, if there are no notes for participants, it may be a good idea to prepare some — perhaps half a dozen sentences outlining the main points. Not only will this inform the participants, it will also offer a safeguard in case the teacher omits some vital information in the verbal briefing. It avoids the teacher being placed in the embarrassing position of having to interrupt the simulation and say, 'Sorry, I forgot to tell you that . . .'

If the teacher does not prepare or present any notes for participants, it is useful, possibly essential, to prepare a checklist of points to make during the briefing. Various home-made documents can also help to explain to the students how the mechanics of the simulation will operate. The teacher might draw up a chart or diagram or map showing time sequences, location of furniture, and so on.

Other types of materials are those which help to make the simulation more efficient or more realistic. They will probably be mentioned in the controller's notes, but even so it is still worthwhile for the teacher to sit back and think what else might be useful. Here are a few examples: clean note pads and sharpened pencils, little flags, name tags, green baize cloth on the boardroom or cabinet office table, telephone, typewriter, pocket calculators, vase of flowers, clock, coffee and biscuits, wastepaper baskets, formal message pads, tape recorder, maps, tray of paper clips, carafe and glasses for water.

Classroom furniture

The question of the geography of the classroom furniture does need to be thought out beforehand. Provided the teacher has already participated in the simulation, there are no real problems about where to put the tables and chairs, but plenty of opportunities to add to the realism.

The starting point can only be the simulation itself — its nature and structure. The requirements can be one single table around which everyone sits or separate tables or areas for different teams or individuals.

If secrecy is important, teams should be as far apart as possible — perhaps one in each corner of the room, or with different rooms of their own, as recommended in some business and foreign affairs simulations.

Some simulations change the pattern of interrelations at different stages, and in this case it is important that the classroom furniture should be changed as well. It may seem a chore to move a few tables around, but it is always useful and sometimes essential that this should be done.

A team should have a base from which to work. This should be allocated by the controller to a suitable spot, bearing in mind the location of other team areas, the controller's own area, and the likely traffic between bases.

Should the simulation involve interviews, furniture arrangements should copy the normal furniture arrangements for interviews. In a public inquiry the chairman and advisers should have one table dominating the room, with the other chairs either facing this table in rows or as cross benches if there are 'for' and 'against' parties.

What should not happen is that the teacher walks into a class in which there are neat rows of desks, simply hands out material and asks the students to form teams and find tables for themselves. This is a prescription for time-wasting and muddle. If some teams should be close to each other, the controller must arrange this; if far apart, this also should be indicated. Maps of the classroom showing which groups sit where at each stage of the simulation can be useful. Such classroom plans not only save time but allow the students to do their own furniture moving.

If the simulation already involves a map of a geographical area on which countries (teams) are represented, the classroom geography might be copied from the map, with neighbouring countries having neighbouring teams. This may cause an espionage problem if the neighbours are hostile. There are several ways of solving this: a screen between the two teams, a strong warning against 'cheating', espionage not being allowed, or the two teams sitting facing away from each other.

The location of the controller's desk should be within easy access of all the participants, but preferably not impeding busy traffic routes.

If the controller has to fumble uncertainly with classroom furniture during the course of the simulation, this is almost as bad as not getting the materials right. The controller who says, 'Well, perhaps we might put this desk here, and you could go over there — you, not you — and then we'll have those materials over here on this table', will not endear himself to students who are in the middle of an exciting and involving experience.

Special provision will have to be made if the simulation includes the media — newspaper, radio or television. Each media organization should have its own base with the necessary equipment.

A studio can be a problem for teachers who are not experienced in this sort of thing. However, a simple and acceptable solution is a tape recorder with a microphone placed on a chair on a table. A coat over the back of the chair placed between the microphone and any likely source of external noise helps to improve the quality of the recording.

Broadcasts should always be live. (Fancy recorded inserts rarely work and take away the personal quality of immediacy.) They should also be done standing up, moving towards and away from the microphone at appropriate moments if more than two people are involved. The alternative of sitting at a table usually introduces a mass of clunks and squeaks as the microphone is pulled in front of the speaker, or chairs are thrust back and forth.

Television broadcasts are much the same as radio broadcasts, except that the viewers should all be at one side of the room and there should be markers to indicate the left and right of the screen to indicate when a speaker moves into or out of camera shot. So all-pervasive is television that this convention is readily accepted by the students. No cameras or technicians are needed.

Masterman (1980) gives some useful examples of how simulations can help in courses concerned with learning about television. Media simulations can be done in style with actual video equipment and with trained and practised operators.

However, a warning note should be sounded about using available electronics. Many a simulation has been brought to a shuddering halt by failure to operate the equipment correctly. Laughter producing incidents can easily arise through failure to spool tape successfully or to cue participants, or edit the videotape. The simulation can grind to an embarrassing halt, rather like a lecture where no one knows how to operate the projector.

If done successfully, of course, there is the advantage of an electronic record of what happened which can be used for demonstration purposes later. But the controller should think carefully before embarking on the use of electronics. Is it necessary? It is desirable? Is it merely a gimmick? Is it worth risking the value of the simulation? Will the actual process of recording the action detract from the realism? With a news conference, of course, it is permissible to accept that it will be recorded and that things could go wrong in setting up the lighting, microphones, and cameras. But if it is a confidential cabinet meeting, or an interview of an applicant for the headship of a comprehensive school, a battery of electronics — and the possibility of hold-ups and failure — may not be worth the effort, and would almost certainly detract from the realism and inhibit the action.

Business and military simulations can be different. If these need computers to work out results, they must be provided.

Timing

The timing of simulations can present problems or involve guesswork as some activities can be completed quicker than expected or continue longer than expected. Again, the teacher's own participatory experience in the simulation is invaluable, not only in assessing how long things will take, but also in drawing up contingency plans in case the timing does not work out.

However, nothing really disastrous can go wrong, provided that no vital activity is started without sufficient time to complete it. If the climax of the simulation is a radio transmission, it will be a shambles if the lunch bell goes when the broadcast is only half completed. Similarly, if a person is to make a set speech, it should not be begun if it cannot reasonably be finished in the time available.

If the simulation contains an activity which has to be treated as a whole and not interrupted, then the timing must be tailored accordingly. Deadlines must be firmly laid down, guillotine procedures established and starting times observed. Deadlines, in particular, must be regarded as sacred. The participants have the role of professionals and they must behave like professionals. If the broadcast is due to start at 2 pm, the participants cannot plead to the controller 'Give us a few more minutes, we are not quite ready.'

Breaks in simulation activity should not be regarded as necessarily undesirable. They give the participants time to think out their strategies and explore their options. A chat over a cup of coffee or an evening at home can help the participant to think himself into the role.

In deciding how to fit the simulation into the time available, consideration should be given to the breaks. It is helpful if the briefing, and perhaps the handing out of some materials or roles, takes place on the day before the action, although contingency planning may be needed in case someone does not turn up or unbriefed students arrive.

Some simulations last for days, weeks or even months. In these cases it is sensible to make sure that teams have adequate resources to carry on in the case of unavoidable absences.

Stages

A simulation which has various stages may present problems in moving smoothly from one stage to the next.

With model-based business simulations each stage may be similar in form to the last one and no great changes need occur. Nevertheless, it is worth considering what should happen when one team finishes its decision-making much earlier than the rest. Do they analyse the data further, do some other task, chat, do crossword puzzles, go for coffee, or snoop on the other teams? The teacher should have some suggestions to put forward in the briefing, and this means thinking about the problem beforehand. If there is a delay in processing the decision-making forms in order to present the results to the teams, it is useful to warn the participants in the briefing, thus avoiding or reducing any unnecessary frustrations during the action. However, there are various devices for reducing or even abolishing such delays. One technique is to time the end of a session to coincide with a normal classroom break — leaving the controller (or clerk) to feed the decisions into the arithmetical model and produce the results in time for the students' return.

A different approach is to speed the calculations. Some simulations contain ingenious devices involving interlocking cardboard circles, allowing instantaneous read-off.

Pocket calculators have suddenly emerged as tremendous time-savers, especially the new models which can be programmed with a formula. Once the formula has been fed in, the actual processing of the decisions is rapid. These new pocket calculators also give flexibility to the simulation designer and the controller, as different manufacturing or trading conditions can be represented by changing the formula and feeding in another programme.

The way to abolish delays altogether is to introduce a time-lag of one stage between decisions and results. Thus, the teams are always lacking the results of the previous stage. At the end of the first stage the participants hand in their decisions, but receive no results at all and go on to the second stage. When they have completed their decision forms for stage two, they receive the results of their stage-one decisions. After the decisions of stage three, they receive the results of stage two, and so on. This arrangement allows the controller or his assistants plenty of time to process the results, since they can do the calculation during the period when the participants are making their decisions for the next stage. The arrangement can be justified on the additional grounds that it is realistic: many business decisions do not produce instantaneous results, therefore predictions and guesswork are needed.

Finally, there is the technique of informational breaks. These are used for instruction, preferably connected with the simulation itself. The students might receive a mini-lecture about data analysis, cost accounting, animal husbandry, or quality control, at an appropriate point, while the controller or his assistants produce the results of the previous round of decision-making.

These techniques about bridging the gap between one stage and another are also appropriate for certain types of non-business simulations. In an international affairs simulation, for example, there might be a mini-lecture on the procedures of the credentials committee or the security council or the commission of inquiry, or whatever is to be the scene of action in the next stage.

Sometimes these informational breaks can be incorporated into the realism of the simulation itself. The controller might take on the part-time role of minister or official giving a news conference, or departmental briefing. If this can be done, it means that the participants might stay within their roles during the informational breaks.

Some operations lend themselves to natural breaks. A simulation in which some participants represent newspaper correspondents or reporters means that there will be a break between the end of their reporting and the publication of their stories. Alternatively, it may be a good idea to have two teams of reporters alternating with reporting and publishing, so ensuring continuous reporting of an event and continuous publication.

When considering the question of stages, the teacher should look at the nature of the activities of the individual participants — what sort of things they have to do and how long it takes.

Re-runs

The possibility of a re-run should be considered before the briefing because it might affect the way the simulation is presented, and also the way the simulation operates. For example, if the students know that there will be, or may be, a re-run, they will have an additional motive for paying attention to what the other participants do, as they themselves may have that role in the re-run.

Re-runs allow the participants to switch roles and find out what it is like being on the other side of the table or the other side of the argument. They enable the participants to have another go. This is important, as students are usually self-critical of their behaviour in simulations, and are often far from satisfied with their actions. Knowing more about the simulation at the end than they did at the beginning, they are often anxious to try it again.

For the controller, the scheduling of a re-run means that there need be no anxiety that certain 'lessons' or opportunities are missed first time through.

However, no simulation should be re-run unless there is a good reason for it. Simulations which have repetitive stages are unlikely candidates for a re-run, unless the arithmetical model or some fundamental condition is changed. Simulations with hidden agendas are unworkable the second time, because the 'answer' was revealed the first time. Equally unsuitable are simulations which are like puzzles with one right answer.

Adapting

Many teachers adapt materials. Making adjustments to suit the classroom conditions is permissible, but it should never be wholesale adaptation and no part of the author's work should ever be copied or reproduced without permission as this is illegal, being an infringement of the copyright laws. In music, for example, taking a few bars of an essential melody may constitute an infringement of copyright. As far as the adaptation of simulations is concerned, the main danger is that the teacher will start the adaptation before the simulation has even been tried out.

Because simulations are so difficult to assess in their packages, it is often the case that the teacher decides a certain document is too complicated, or a role is unnecessary, or a lot of extra facts have to be fed in before the participants can start. The teacher can ruin an excellent simulation in this way. Key bits which are parts of the checks and balances can be removed or rendered ineffective. For example, the simulation may contain a boring, jargon-filled document about health hazards or unemployment benefits, and the teacher may decide not to hand it out. Yet the simulation within itself may generate a need to tackle such a document for some specific purpose — to report on it, to interview someone about it, to use it in an argument as evidence, to ask for its amendment, etc. In the action part of simulations, documents are rarely read dispassionately.

Generally speaking, the more experienced the teacher is in using simulations, the more willing he is to use the simulation according to the recommendations of the author. At least he will for the first try. Later on, any adaptation will be based on experience not guesswork. The sort of adaptations that are really useful are those based on simulation philosophy; that is, adaptations which try to make it run more smoothly, increase student responsibility, and improve realism.

For example, what can be done if the simulation works reasonably well but has a bulky scenario which has to be digested before the action starts? There are several possibilities. The information might be fed into the simulation in stages. Some of it might be incorporated into the individual profiles. The mode of presentation can be varied — a letter in students' pigeon-holes, a tape recording, even a telephone call. If the teacher sifts through the information, some of it might be separated into a 'library' where it can be delved into and skimmed through according to the various strategies and interests of the individual participants.

Should the simulation seem to stick at particular points, with no one showing initiative, the teacher can see if it needs a drop of oil — perhaps the introduction of the role of journalist, or of some provocative document.

Similarly, the simulation may seem to be less effective than it might be because of part-time or passive roles. For instance, can a jury be dispensed with? Is it necessary to have a United Nations Secretary-General — or could the controller take on this role on odd occasions? Is a banker an essential role or could it be left to the participants to manage their own transactions? Is the representative of the local authority such a peripheral role that for most of the time he stands around watching?

Practice and experience are necessary for distinguishing useful waiting from boring or frustrating waiting. A person waiting to appear before an appointments board may be sitting around and, to all appearances, doing nothing. Yet mentally he may be very active rehearsing his strategies and predicting likely questions. After the interview, he may also sit, slightly more relaxed, but still engaged in a mental inquest on what happened.

Waiting can be alert, watchful and strategic, with the potentiality for action, and the mental recording of who is doing what and why. This is not the same as looking blankly out of the window, reading a comic or doing a crossword, or telling funny stories.

There can be several reasons for adding to a simulation. The addition of 'public opinion' — perhaps journalists — to a foreign affairs simulation can add more realism, introduce a constraint to the use of power by statesmen, and provide an impetus to diplomatic moves. The controller can add facts — perhaps in the form of leaflets in a 'library' — in order to give more background data and ammunition for argument. If the main aim is practice in communication skills, it might be possible to add a sequence involving a news conference, public meeting, or parliamentary debate.

However, adaptation of this sort should not be undertaken without a good deal of thought and a reasonable amount of experience with simulations. Really major adaptations should not be attempted: it is better to buy a more suitable simulation.

A special case for adaptation is in the teaching of English as a foreign language. But even here, the teacher should give the simulation an unadapted run first, as it is easy to underestimate the ability and motivation of students involved in a simulation.

Briefing

Briefing is easy, providing it is based on personal participation by the teacher and on careful preparation.

If the students are unused to taking part in simulations, it is useful to spend some time explaining what they are and what they are not. The key points to emphasize are the extent of the powers, duties and responsibilities of the participants, and also the dividing line between reality within the simulation and the fictional background outside. Even if the students have taken part in simulations before, it is advisable to make sure they are aware of these points.

With careful preparation, the controller will enter the briefing well primed with explanatory notes, diagrams, maps, timetables, deadlines, or whatever else is necessary.

If the simulation is simple, the controller may be the only person in charge. But with a longer or more complicated simulation it is advisable to have an assistant or even a 'control team'. Some teachers who are familiar with simulations prefer giving one or two of the students the opportunity to be controller as it gives them practice in organization.

Naturally, the briefing should contain no hints or nudges about policy decisions. Therefore, for the controller, the briefing is a very practical session — a checklist of items to be dealt with, points to be made and queries answered. Should the simulation have several stages, parts of the briefing can be left until later and the information given immediately before the stage in which it is needed. The more thorough the briefing, the less likelihood of unexpected events arising which could interrupt the simulation or knock it off course.

The controller should be cautious about giving too much information to the participants during the briefing. A student who is placed in the management team of Blogsville Ropes Ltd might ask, 'Does it manufacture all types of ropes?' If the answers to the questions are available in the documents, it is probably best for the controller to decline to answer. In most simulations, it is one of the functions of the participants to find out the facts. If the briefing opens the door to this sort of factual question, other participants can start asking similar questions and the controller can end up by spoonfeeding information to the students. Also, the factual information provided by the controller in this off-the-cuff manner could be distorted or misleading. It is better to let the students find it out for themselves from the documents.

This will, of course, leave some students dissatisfied and feeling that they do not know enough about the situation. But the controller can explain that adequate information will be provided when they receive the materials and that the briefing is concerned with the mechanics of the simulation only.

Action

Controlling some simulations is as easy as rolling a ball down a gentle slope. Even with more complicated or lengthy simulations, the control problem is simple providing the simulation has been well prepared and briefed. The reason for this is that the action tends to look after itself. It has its own power, its own catalysts and its own initiatives. The problem with a good simulation is not to get it moving but to get it to stop.

In a simulation the controller is generally in a good position to observe student behaviour, including fact learning, strategies, decision-making, problem-solving, and the use of language and communication skills — to say nothing of enjoyment. The controller should have a notebook handy to jot down various points as they occur — possibly for use during the de-briefing.

The vital task for the controller during the action is to make absolutely sure that the right materials are available as and when they are required. Once the simulation is under way, the controller should check the materials, which might have been shown to the students during the briefing, and make sure they are in the right order and grouping for handing out or for availability. In some simulations this is unnecessary as all the materials are given to the participants at the start, but any simulation which involves feeding in materials from time to time requires close monitoring by the controller.

In addition to the question of materials, the controller should also prepare in advance for changes in classroom furniture, timetables or role changes which might be part of the structure of the simulation.

With a short simulation — lasting, say, less than an hour — no intervention by the controller should be required assuming, of course, that it has been adequately prepared and briefed. But with longer simulations some minor adjustments in the machinery might become desirable or necessary and, very occasionally, a major change is required. In these circumstances, the problem for the controller is whether or not to intervene and, if so, how and when. Effective intervention may require skill, experience and imagination.

When intervening, the controller should have two objectives — to interfere as little as possible with the smooth running of the simulation and to select a cover story which fits in with the simulation itself.

Suppose that during a simulation involving interviews or a public inquiry, the controller notices that the members of the board or panel say, 'Now then, Mr ... er ... er'. This may indicate either that the person interviewed does not have a clearly written name tag, or the member on the board has no list of names, or both. At the next suitable break the controller, adopting the guise of messenger boy, usher or whatever, provides the necessary name identification material to whoever requires it — perhaps even apologizing and saying that the town clerk's department was responsible for the omission.

With a medium-length simulation, lasting half a day or a day, or a longer simulation which goes on for more than a day, the chances increase of the need for a change in roles. A student may go sick, or may not turn up, thus creating a serious problem. If the simulation consists of teams, the missing member may not make much difference; but if the missing person happens to be the prime minister possessing secret information, it will probably be necessary to halt the simulation in order to make suitable adjustments.

Very occasionally a change of role is necessary because of an internal coup. The prime minister may be overthrown because there was a basic conflict between his 'sell-out' policy and that of public opinion as represented by his country's media. The control team may have decided that the prime minister was wrong and should be deposed, or 'forced to resign', or whatever is plausible. Any major change of this nature cannot be dealt with peremptorily. There will almost certainly have to be a break in the simulation until the roles are changed, someone else becomes prime minister and the ex-prime minister takes another role.

A role change would be required in a local affairs simulation if the chairman of the council finished up on the wrong end of a vote of confidence because of repeatedly favouring one group. In this case, the intervention by the controller might be quite unnecessary, since the participants could deal with the role change within the confines of the action itself — there being no need to hypothesize a general election or military takeover in order to move a person from one chair to another.

It is rarely the case that the unexpected arises unexpectedly. There are usually warning signals and the controller should watch out for them. In this way, drastic intervention can often be avoided by taking minor remedial action. Even if the major disruption occurs the controller will have had time to work out some contingency plans.

Suppose the controller notices that a participant has nothing to do and is looking bored. The question is whether this is slight and temporary or whether something has to be done about it. If something has to be done, it should be within the plausible parameters of the simulation. A note can be handed to the participant from the local council, the editor, the managing director, the shop steward, or the prime minister, asking him to help X or Y.

Should any serious misbehaviour occur, probably the best thing is for the controller to send the person concerned a note asking him to come and take a telephone call, or come and see the cabinet secretary, military commander, etc. Having extracted the trouble-maker the controller can find out what is the matter. It might have nothing to do with the simulation, or it might be a misunderstanding, or a failure of the simulation to allow the participant a full role.

Having found out what is the cause of the trouble, the controller can take whatever action seems appropriate.

It is much easier for the controller to deal with behavioural problems within a simulation than in normal classroom teaching since there is no escalation of personal antagonism between student and teacher. The controller, by the nature of the job, is not eyeball to eyeball with the students and consequently is a detached and impartial authority.

The main danger in the action part of a simulation is not misbehaviour so much as inappropriate behaviour, and this may be caused by the failure of the controller to explain clearly enough what can and cannot be done in a simulation. As has been emphasized earlier, the participants must be told to accept their function: they are businessmen, town councillors, or world leaders, not magicians, gods or saboteurs.

Supposing, for example, in a history simulation a Saxon king announces that he has moved his army 100 miles overnight. What happens next? If the simulation has been well briefed, then one or more of the other participants will challenge the decision. One challenge would be to point to some of the documents or to commonsense and say that 100 miles is an impossible distance to cover in such conditions in such a short time. But a more important challenge would be to say that an order is only an order, and a decision is only a decision, and that the 'facts' outside the room depend on the controller. Even if it was an order for an overnight journey of only one mile, it would still be up to the controller to decide whether the order was carried out, whether it was effective, and what other consequences resulted from the order. Assume, however, that the briefing was somewhat inadequate, or that the other participants were not sharp enough to realize that the Saxon king had turned into a Saxon magician, what should the controller do about the inappropriate behaviour?

One way is to stop the simulation and explain to everybody that this sort of thing is not allowed; but this has the disadvantage of disrupting the flow of the simulation. Another way is for the controller to send a written message to the Saxon king saying, 'My lord king, your army is foraging for food. It is not possible to march until tomorrow.' This should hold the situation until the next break, during which the controller can explain the difference between inside decisions and outside facts.

In ordinary teaching it is customary and efficacious for the instructor to step in to correct mistakes of fact. It is also usual to go further than this and offer advice and information about non-factual matters — questions of ethics, values and opinions. For a teacher, these interventions become second nature and habitual. During a simulation there may be an intensely strong and perhaps overwhelming desire to step in, interrupt the action, and convey the correct information or the useful piece of advice. The intellectual justification for this sort of intervention is that if wrong information goes uncorrected it may be learned and reinforced by repetition, and that the controller should provide the correct facts in order that they may be put to practical use, tried out, tested and learned while participatory interest is high.

Some authors go further than this. They argue that the controller should ensure that the discussion is relevant, that each person has a fair opportunity to contribute, and that valuable points are discussed in plenary session during the action rather than just within one group.

The trouble with this sort of intervention is that it will kill a simulation stone dead, leaving only an instructor-controlled exercise. There is nothing wrong with instructor-controlled exercises, but they should be advertised as such, not presented under the guise of being a simulation, with participant responsibility conferred and then taken away.

On the strongest point of the argument — correct facts — it is necessary to distinguish between the fictitious 'facts' — the Blogsville Company's production costs, the Ruritanian defence treaty, etc — and the non-fictitious facts, the real facts of the outside world which may impinge on the action. In the first case, no 'facts' will have been learned incorrectly; the only incorrect learning is the fictions. As in everyday life people make errors, read documents incorrectly, and so on, and in a simulation they should pay the normal penalties for carelessness, and not have the controller protecting them from the folly of their ways.

On the question of real 'facts' being given incorrectly, these are either in the documents or they are not. If they are in the documents, the author or publisher is in error or the facts are out of date. The controller should have done something about it before the simulation began. If the 'facts' are not in the documents, they are nothing but allegations made by individual participants and should, as in everyday life, go uncorrected if they pass unnoticed. The controller can, of course, make a note of the incorrect allegations and point these out in the de-briefing, but he should never interrupt a simulation simply to correct facts. Interventions of this sort disrupt the flow, diminish student responsibility and ownership, constrain behaviour for fear of 'getting it wrong', cause resentment, and open the door to an attitude of 'Now we are back in school again.'

The only case in which intervention regarding 'facts' is justified is when the participants get it wrong in a big way, and are under such a serious delusion that the simulation itself is imperilled. In this case, the best thing is for the controller to break off the simulation at a convenient point and correct the misunderstanding by whatever method seems the most plausible, depending on the nature of the simulation.

Even when using a simulation to teach English as a second language, the teacher is well advised not to intervene to correct linguistic and grammatical errors. The teacher has to ask the question whether it is to be a simulation or a linguistic exercise, whether the aim is error-free language or successful communication.

In this context Kerr (1977) says:

> In the course of a simulation, the teacher may be tempted to intervene when mistakes are made, or even to introduce brief spells of remedial teaching. In general, this is unsatisfactory from several points of view; the student being corrected finds that his train of thought has been interrupted, while the teacher will probably find that the students are not paying full attention to his explanations, but are anxious to proceed with the simulation. Experience has shown that it is better for the teacher to sit in the background with a notepad, jotting down errors as they occur. It is usually convenient to timetable a remedial teaching lesson (immediately after the simulation ends) in which important mistakes can be discussed and remedial practice takes place. Another possibility is to tape record all or part of the simulation and play back the recording to the students afterwards, inviting them to identify their own mistakes as they listen.

This advice about recording a simulation can be useful in contexts other than learning English as a foreign language. Many people speak badly or mumble or fail to put forward their ideas in a way that can be understood. Even reading aloud is a dying art, with the person concentrating so hard on looking at the black print that each word comes out but the meaning stays behind. A tape recording helps to illustrate the point. On the other hand, there are dangers that the intrusion of recording apparatus could inhibit the participants. If tape recorders are to be used, they should be used as often as possible so that the participants get used to them and forget about them.

De-briefing

In the follow-up discussion or de-briefing, the controller returns to the role of teacher or instructor. The transition need not be abrupt, however, since it may be a good idea to allow one (or two) of the participants to take the chair, particularly if the simulation itself involved this sort of function. In this case, the teacher's contribution would have the same status in procedural terms as that of the students; for example, in giving an account of what the mechanical problems of the simulation were and how the controller tackled them.

As a general pattern it is useful to go round the table and have the participants explain their own parts in the simulation — what they saw as the nature of the problems and how they dealt with them. Each participant would contribute to the sum of knowledge, which would be particularly useful if there was a divergence in roles and functions. It also adds to the practice in communication skills to be able to explain what one did and why.

The second stage, after everyone has had his say (without comments or discussion), involves general discussion. The obvious way into this general debate is by looking at the immediate specific questions relating to the outcome — the inquest on the result. But this should not be allowed to degenerate into a re-run of the arguments used within the simulation.

In other words, the de-briefing should move fairly rapidly from the particular to the general. The real value in a simulation will be in the transfer of knowledge and experience to other situations in the future. This is likely to involve general principles — how did the groups organize themselves and was the organization effective? What alternatives were there? Did the group or individuals explore the options, analyse the nature of the situation in which they found themselves, and plan accordingly? How effective was the communication? Were the language and behaviour suitable and appropriate? What lessons did the participants learn? Would they act differently when faced with such a situation in the future?

The controller may also be keen to get the reactions of the students to the simulation itself — its materials, mechanics and general situation. Here it may be necessary to interpret various remarks. If a student says, 'That was great fun', the comment can mean that the simulation was not much else. If someone says, 'I lost my bit of paper' instead of 'I lost my housing document', this may indicate a lack of realism in the materials. 'I got bored during the last part' could mean that the participant had a part-time or passive role, or that the controller had not presented the simulation in a satisfactory manner.

Should the students say, 'We'd like to have another go', it could mean that the teacher should have considered the possibilities of a re-run, and reached a decision on whether to have one or not or leave it to a vote. Naturally, if there is a re-run, it is advisable to postpone the de-briefing or to curtail it, otherwise too many hints and pieces of advice may be given. But this is a matter for on-the-spot judgement.

In a simulation with a hidden agenda a thorough de-briefing is usually essential. This is particularly true in behavioural simulations. These simulations are likely to stir up emotions which can last much longer than the simulation itself. Some participants may feel exposed or humiliated. Gaining insights into one's character can be an abrasive experience. Some participants might feel that they have not gained insights, but have been cheated or manipulated into expressing attitudes, views or emotions which are contrary to their characters or personalities. Little advice can be given to the controller in such circumstances, since the problems and likely outcome will already have been anticipated and will probably have been the reason for presenting the simulation in the first place. The controller should at least make quite sure that there is ample time in the de-briefing to explain why the simulation had a hidden agenda and what it was supposed to reveal.

In general, a de-briefing should not be regarded as an optional extra or something to be dismissed in a sentence or two. As will be seen in the next chapter, it can be useful for the teacher to organize an occasional in-depth de-briefing of two or three of the participants.

Chapter 5

Assessment

Aims of assessment

Research and evaluation is an area of controversy in simulation literature, as indeed it is in education generally. The debate is usually on whether the methodology and design of the experiment is appropriate and whether the findings justify the conclusions.

But from the point of view of the teacher, there is nothing forbidden, or even forbidding, about assessment in relation to simulations.

Assessment does not imply some grandiose pie-in-the-sky research project to test the hypothesis that 'simulations produce greater gains in critical thinking, decision-making, and problem-solving than do other learning methods', or some such similar generalization. Many authors have pointed to the unavoidable difficulties of such attempts. Davison and Gordon (1978) point out that no evaluatory instruments can readily encompass the many different dimensions of behaviour and experience involved, and Twelker (1977) emphasizes the problems caused by the great differences between individual simulations and between the conditions in which they are used. In the jargon of research, there are bound to be a great many uncontrolled variables.

The teacher, therefore, should aim only for what is functional and practical. The idea should simply be to learn something — something about the individual simulation, the participants, and also the teacher's own thoughts and behaviour.

It is a pity that virtually all writers on simulations talk about assessment and evaluation exclusively from the point of view of assessing a simulation. The only criterion seems to be whether the simulation is useful, appropriate, stimulating, etc.

For the teacher, this is a restrictive way of thinking about assessment since it is only half the picture. Just as people can assess simulations, so can simulations be used to assess people. As well as examining simulations, simulations can be examinations. Organizations which have been using simulations the longest — the armed forces and the higher levels of the civil service — use some simulations for the specific purpose of testing.

At staff colleges, on courses at country houses and in the interiors of ministries and large organizations, simulations are used, together with puzzles, problems, case studies, discussions, etc, as devices not just for training, but also for assessing the participants.

Once the instructor, tutor, consultant or teacher becomes familiar with a particular puzzle, simulation, etc, it can be used to increasing effect for assessment. Practice and experience are needed, as in any other field. There is nothing mysterious about it, nothing really difficult, as every teacher is a professional assessor. Simulations may be unfamiliar territory, but any teacher who has read the first four chapters in this book will have a good idea of the sort of things to look for.

Language and communications

The report of the Bullock Committee, *A Language for Life* (1975), refers to an experiment conducted by the Southern Regional Examinations Board in association with the University of Southampton to study the examining on a large scale of oral English. Some 450 candidates were divided into four groups, each of which took a different form of oral examination:

1. Reading a passage and talking with the examiner
2. Making a short speech or lecture and answering questions
3. Talking to the examiner about a diagram previously studied
4. Participation in group discussion.

These four forms of examination were supported by tape recordings and by an assessment of the candidate's spoken English by his teachers.

The conclusions, which were given tentatively, suggested that the most successful and also the most natural method was the second one — making a short speech and answering questions. Method 3 — talking about a diagram — led to the cultivation of 'civilized conversation', while Method 4 measured ability which was unrevealed in the normal classroom situation. The research workers said that habits in spoken English could be 'sharpened, enriched and disciplined by intellectual and sensitive attention in the classroom and syllabus' and that this attention could be focused on the problems by the Certificate of Secondary Education examination, a subject-based examination usually taken at 16+ and administered by regional boards in England and Wales.

Following this experiment, all the CSE regional boards have introduced examinations in oral English, and certain boards include in the course such optional activities as improvised and scripted drama, debates, tape recorded interviews and oral comprehension.

It is a pity that simulation was not one of the methods used in the experiment, since the Bullock Committee, while welcoming the development, makes the point that the forms used are artificial in varying degrees. It quotes one teacher as saying:

> One of my most curious activities each year is helping to conduct a test in conversational English under the direction of my CSE examination board. In this the candidate comes into my room to conduct a conversation with me and another 'examiner'. Thereafter it is our task solemnly to award him a mark out of ten. Nothing less like a genuine conversational situation could be imagined . . . Talk, as a medium of social intercourse, cannot be reduced to the level of an examination mark.

The Bullock Report says it is only fair to point out that some examination boards have gone out of their way to eliminate this artificiality, but the Report recognizes that it is a fundamental dilemma. It is to be hoped that the examination boards will look into the possibilities of using simulations.

Teachers who devise their own Mode 3 syllabuses for the CSE examinations are not debarred from using simulations as tests since they could be regarded as part and parcel of the continuous assessment made by the class teacher which adds up to an examination result.

But apart from the question of any formal or external examinations, any teacher can consider the use of simulations as a tool of assessment in language and communication skills. There is no reason why a teacher or group of teachers should not shop around for suitable short, balanced and talk-oriented simulations. The teacher would become familiar with them if they were used as tests. They could be used at the end of the course, or at the beginning, or at both the beginning and the end in order to assess progress. The more they were used and repeated in similar circumstances, the easier it would become to compare one individual with another, and one group with another.

Assessing behaviour

Some organizations use simulations and role-play exercises as tests not only of language and communications but also of behaviour.

Contact between a member of an organization and someone outside involves behaviour, and the organization is usually very concerned to see that the behaviour is appropriate and effective. Salesmen, policemen, shop assistants, interviewers, airline hostesses, and others are all required to conform to certain standards of behaviour. In certain jobs and professions, both appointments and promotions depend a great deal on behaviour.

Sometimes the assessment is recorded formally. For example, if the simulation involves an interview of some sort (appointments board, personal interview, media interview) there may be a standard form for the instructor, tutor, consultant, etc to record the result. See below, for example.

Criterion	Standard achieved			Remarks
	Above average	Average	Below average	
Posture and deportment				
Clarity of expression				
Confidence				
etc				

But whether the assessment is formal or informal, the main question to be asked is whether the behaviour is appropriate to the circumstances. This question can be asked about any participant in any simulation. Depending on the participant's function and the specific job and circumstances, the question can be divided up into various parts. For example:

— Ability to make a point
— Ability to stick to the point
— Ability to search for options
— Ability to work out possible consequences
— Degree of courteousness, sympathy, understanding, honesty, diplomacy, etc.

These ideas can form the basis of a teacher's assessment of the behaviour of participants during a simulation.

In-depth interviews

Assessment can be retrospective. As well as observing what happens during a simulation, the teacher can use in-depth interviews after the simulation to assess both the participants and the simulation itself.

This can be regarded as a supplement to the de-briefing. It can cover similar ground, but whereas the de-briefing is often generalized, the interview can get down to the particular and the individual. Since de-briefing sessions often tend to be on the skimpy side, the interview can be a highly valuable tool of assessment and learning. Its value lies in revealing what otherwise might remain unknown or obscure.

In-depth interviewing is rather like beachcombing. All sorts of interesting and curious things come to light, some of them most unexpected. It can reveal insights into attitudes, events and motives which range much wider than the preceding simulation event.

Perhaps the first thing to be decided is who should be the interviewer. Should it be the teacher, or a colleague, or a counsellor, or a consultant, or an outsider? Is it better to have someone who knows the participants well, but who has had little experience in interviewing, or is it better to select someone who is an expert interviewer, but who does not know the participants? Like many other questions, there is no 'right' answer to this one, but whichever way is decided on could be changed on subsequent occasions in order to explore the effectiveness of the various options. Important ingredients are time, privacy, and a tape recorder.

As for the interviewing technique, there are no easy rules, and each person has an individual style and set of interests. Since there are few examples in simulation literature of this sort of interviewing, here are excerpts from interviews conducted by the author after students at a London college of further education had participated in the author's *Nine Graded Simulations*.

The first interview was with a 16-year-old Asian girl, and was remarkable because the author had formed the impression on watching her in the simulations that she was confident, a good talker, lively, likeable, and had lots of friends.

Q: Pretend I know nothing about simulations at all and I say to you: What are these things you've been doing? Are they any good? What happens? What do you feel about them?
A: Well, I think they are very good actually, because they tell you about life. They actually teach you something you would never know about the society in which you live.
Q: Do you think they help you to become more open-minded or more dogmatic?
A: They help me personally to become very open-minded. They help me very much. Because, before I did these simulations, I never had any interviews or anything, and it was the first time I did any of

these things. And now I think, in future when I go for any
interviews or when I do anything, I'll know actually what is
happening, and I'll be more confident. Before I did these
simulations I didn't have any confidence at all.

Q: You felt you were not doing justice to yourself?

A: Yes.

Q: You felt you'd got ideas, but somehow something stopped you
bringing them out?

A: Yes.

Q: Apart from learning about reality and getting more confidence,
did you actually like them?

A: Yes, I liked the one on the local radio broadcast. I was one of
the broadcasters, and being on the air really is a nice feeling. It
was the first time I had heard myself speak, and I pretended to
be a news broadcaster, and I had a feeling I was an important
person and that all the world was listening to me, and I was
proud of it.

Q: Do you talk about it afterwards to people who have not taken
part in it?

A: Yes, I do, I talk to my friends about it.

Q: What do you tell them?

A: I tell them everything we have done in the simulations.

Q: Do they say: So what, what a waste of time, I wouldn't lil
do that?

A: Well, some of them do, but the others are really very interested.
They ask me questions about it afterwards, you know. Like what
happened, or what we did in the simulation.

Q: Do you feel they have actually helped you to talk?

A: It does help me a lot, because before I did any simulations I
always used to stammer a lot. And I s...s...still d...d...do
[*stammering badly*] but now I've gained confidence in speaking
to the public [*recovering from stammer*] and other people, and
in fact it helps me a lot.

Q: What about relationships with your classmates? If you hadn't
taken part in the simulations, do you think your attitude
towards them would have been any different from what it is
now?

A: In our class we have students from everywhere, all over the
world. When we first came I was too shy to speak to any of
them. But after the simulations, I used to, you know, have
confidence in speaking to them. I'd have something to talk to
them about. There wasn't anything to hold me back, or
anything. I could easily talk to them.

Q: Why had you been held back before, do you think?

A: I'd never spoken to anyone before, I mean.

Q: Why not?

A: Well, I didn't have anything to talk about, and I didn't feel I was confident in what I said to talk to them. I used to see people as strangers but now, after talking to them, I consider them as friends. You know, I'm at ease with them.

Q: And you feel this would not have happened if you hadn't had the simulations?

A: No, I would be sort of . . . I would keep myself to myself all the time and I wouldn't have had confidence to talk to anyone.

Q: These simulations were all very different. Wouldn't it have been better if they had all been on the same subject?

A: Oh no, because you need to know about everything.

Q: Each one's a new situation. Doesn't it make it more difficult to jump about between being a radio journalist, and a candidate, and a . . .

A: It makes it more difficult but, on the other hand, it makes it very interesting.

Q: Because it's new?

A: Yes.

Q: How do you feel about entering a new simulation? Don't you feel nervous and shy, not wanting to take part?

A: Oh yes, I had this feeling in the first few when we started. I wanted to work in a group and did not want to do anything by myself. But then after proving myself successfully, I was a confident person and I was prepared to do it on my own and everything.

The reason the author repeated some questions was sheer incredulity when the girl began to stammer and talk about never having any friends at college before the simulations began. This seemed so contrary to her behaviour during the simulations, when there had been no trace of a stammer or shyness.

The next interview was with a very different subject — British, male, aged 27, and by far the oldest student in the class. This is how it began:

Q: Did you learn anything as the simulations progressed? Anything that you know now which would have helped improve your performance?

A: Yes, I think so. I've learned a hell of a lot about things in general. I'm older than most people in the class, and I thought I had a bit more of an insight into life, but I've learned a hell of a lot from these simulations about how things work, that's the principal thing. If I can make a comment I think that simulations are really good if they are taught at the right stage in schools, or if they are given at the right age, because so much in education is that you sit there . . . When I was at school you sat there like some silly sponge soaking up all this information without being able to give anything out. And the one thing these really do is give you a chance to put something up, and you learn something from that.

Q: Instead of being a sponge, you're a human being?

A: Yes, exactly. It redressed the balance, so to speak. It's all very well knowing about Christopher Columbus, but it's nice to know something about life and how things work. And these simulations, they give you situations, which you think about afterwards, and you think, well, that probably is what happens in real life. There are situations like this, and you learn things from this and you learn how things work.

Q: Do you think you have learned more about your colleagues than if you hadn't had the simulations?

A: Oh yes, particularly where there is competition, as in PROPERTY TRIAL. You learn about the ones who push and the ones who don't, and that's where personalities come in, and it's very interesting and one couldn't help but be affected by this. You see more than you knew, or thought you knew, but you discovered some were pushing and some weren't and some were plain foolish, and that's how they fit into life probably.

Q: I suppose in any group you have some sort of pecking order: that is to say, somebody is more or less top dog, and somebody hardly ever says anything. Do you think these simulations have at all altered the pecking order?

A: I think that what it does show is that nobody is the top dog in any particular situation, and that what it does show is that everybody has something to contribute. In some simulation one person will be top dog, and in another simulation another person will be top dog, depending on the particular personality and talents, or whatever.

Q: Do you talk about them with your colleagues out of class?

A: Yes, definitely. On Friday we go into physics afterwards, and we are still going over them, and it just takes a couple of minutes to calm us down so that we can get on with physics. Actually, it seems very mundane to come out of simulations, and you're all bubbling and you had an argument with somebody over something and you've then got to sit down and become silly sponges again, soaking up all this muck. It's sad, but that's education.

Apart from revealing individual attitudes and ideas, these two
examples were with experienced participants. Each had taken part
in at least half a dozen of the author's *Nine Graded Simulations*.
Interviews with one-shot participants tend to be less emphatic, and
possibly less articulate, since the questions may cover unfamiliar
ground. But these particular participants had obviously already
talked a great deal about the simulations among themselves. They
knew what they wanted to say, and they said it. In this respect they
were typical of all the other members of this class who were
interviewed by the author.

Moreover, none of them seemed particularly self-conscious or
inhibited during the interviews. For example:

Q: Do the simulations help you to talk — to talk in a variety of
   different situations?
A: Well, I'm a very impulsive sort of person, and just talk on the
   spur of the moment, and what I've learned is that you've got to
   be very controlled, you've got to think about what you are
   going to say before you act, you know. You've got to think
   a few minutes before you say anything, before what you actually
   say. Because I usually say what comes out on the spur of the
   moment . . . it comes out in a few seconds . . . and it's taught
   you to be calm.

Self-assessment

Simulations can change teachers as well as students. They learn new
things. Their attitudes towards the participants, towards the
educational institution, and towards teaching in general can be
profoundly affected. They can be stimulated, provoked, challenged,
and jerked into new interests and new enthusiasms. If nothing else,
simulations can be talked about in the staffroom.

However trivial or however profound, it is worth the teacher doing
a little internal monitoring from time to time.

The question should not be simply, 'What have I learnt?' but 'What
did I do?', 'What has happened?', 'Was it unexpected?', 'Should I
follow it up?'

Questionnaires

Stress has been put on the value of using simulations to assess the
people involved. This is often completely overlooked, yet it should
be part and parcel of the teacher's general assessment of the students.

It is no more difficult than assessing discussions, projects, or informal drama and may be easier, as the teacher's role of controller facilitates impartial observation.

The other side of the coin — assessing whether a simulation works — is what the rest of the book has already covered. Since a simulation is best thought of as an event, its assessment must be seen in the light of what happens in practice, and this depends on the quality of the simulation design, the effectiveness of the teacher's preparations, and on whether the simulation is appropriate for the particular students and the classroom conditions.

Questionnaires are a popular tool for simulation assessment. Usually they are tailor-made by the teacher (or research worker or author) to fit a particular simulation. They can be used for small groups or for much larger experiments. They can be used after the simulation has taken place or both before and after.

There are problems. Bloomer (1974) remarks: 'Questionnaires are prone to the danger that the teacher discovers not what has been learnt, nor even what the pupils thought they had learnt, but only what the teacher would like them to have learnt.'

This touches on one of the main problems — what questions to ask? If the teacher limited the questions to the main objectives for introducing the simulation in the first place, the answers will similarly be limited and the questionnaire will not reveal whether any non-specified objectives were achieved.

It is useful, therefore, for the questionnaire to cover skills as well as facts, emotions as well as subjects, and behaviour as well as learning. Fishing with open-ended questions is revealing: 'One important thing that I found in taking part in the simulation was . . .'

However, it is preferable to put the open-ended questions first in the questionnaire. If they are added at the end of a list of factual questions, the participant may simply trot out another fact. Here are some sample questions:

> The thing that surprised me was . . .
> Comment about anything that mattered to you as a person . . .
> How did your talking help your thinking?
> How did you behave?
> Compared with your objectives, did you find what you did satisfactory?
> Was the decision-making in your team democratic?
> Did you learn anything about being diplomatic?

Would you have liked more time for any of the parts in the
simulation?
Did you think it gave you useful practice in (. . . . . .)?
How could you have done better?
Did you consider ethics or only material values?
Would you have introduced the simulation any differently?

If the questionnaire is to be used for statistical purposes, some
questions must have quantifiable answers. They can be

Yes ☐   No ☐

or

Yes				No

or

Very enjoyable						Very miserable

Statistics derived from such questionnaires can be not only
misleading, but sometimes the exact opposite of what they are
assumed to indicate.

An illustration of this is the figure of 3.6 per cent of participants
who listed themselves as being 'totally disinterested' after a large-
scale current affairs simulation which lasted for two days at a
polytechnic. The implication was that the 3.6 per cent were
disinterested in current affairs or the simulation or both. It is the
sort of statistic that causes authors to say, 'Yes, we know that there
is usually a small percentage of participants who don't like
simulations, and here's another example of this.'

Yet on questioning the authors of the research report, it turned out
that the participants who had graded themselves as 'totally
disinterested' were all members of the team representing the
Advisory, Conciliation and Arbitration Service. No other team
called them in. They were unemployed. They had sat around for
two days doing nothing except watch the other participants enjoy
themselves. They were not protesting against participation in the
simulation: they were protesting against non-participation.

Questionnaires should not be used in isolation from other

assessments and should be supported by an adequate description of what actually happened in the simulation in question.

Two other points can be made. First, it is better to have a questionnaire covering two or more simulations than just one. With one simulation the questions tend to float about in the air. What does 'enjoyable' mean? Enjoyable compared with what? With another simulation, with a lesson in mathematics, with a favourite television programme?

Words like 'useful', 'interesting', 'valuable' derive their meaning from comparisons, and if comparisons are not specified, the whole operation takes on a random flavour with many participants deciding to play cautiously and place the tick in the middle of the range of values.

Second, a questionnaire requires to be interpreted in the light of the previous experience of the participants. If they have never taken part in a simulation before their replies to questions may include an extra element of uncertainty, misunderstanding and unfamiliarity because the first simulation is usually the most difficult one.

If the object is to assess simulation X, it is a virtual waste of time to question inexperienced participants. One does not try to assess the value of a specific novel or play by questioning people who have had no previous experience of novel-reading or play-going.

Ethics

Ethical values often find their way into assessments and simulations are no exception. The question of ethics has already been mentioned in relation to behavioural simulations on such controversial issues as race, sex and power. Teachers are understandably anxious that their pupils or students should not be learning the wrong things and acquiring wrong values. The conclusion in Chapter 3 on choosing simulations (in the section on behaviour) was that if it was a genuine simulation and not an informal drama, there would be no implantation of values: it would be up to the participants to behave according to their own ethical standards in the situation in which they found themselves.

The sharpest attacks on simulations, however, are usually related to competition. There seem to be two main causes of concern about the dangers: one is the desire to win to the exclusion of anything else, and the other is that the values are often exclusively materialistic.

Ravensdale (1978) points out that game participants are referred to as 'competitors' and certainly many of the educational games and simulations currently available to teachers are highly competitive. He adds: 'Such games can and do create tensions, show-downs and often fierce competition. That these factors produce a state of adrenalin flow is not surprising; that this extreme attitude can always be useful is questionable.'

Abt (1968) says that 'games of skill have the possible educational disadvantages of discouraging slow learners, dramatizing student inequalities, and feeding the conceit of the skilful.'

Zuckerman and Horn (1973) say that if normal business operations required attention to environmental retrogression, depleting resources and a rapidly deteriorating quality of life, then business simulations would reflect that need. They add:

> Paradoxically, if a simulation were designed to include such constraints, it would be a 'bad' training experience, for participants would be trained in methods of thinking and setting decision priorities which would put them at a disadvantage when it was time for them to compete in the real world. However, it would be foolish to condemn the class of business simulations for this significant lack; they are doing their job, which is training.

And Zuckerman and Horn express the hope that some of the simulation designers, particularly colleges and universities, will 'begin to take an interest in education as well'.

But to be fair to simulation designers, there are plenty of simulations about human values and the environment. Also, plenty of simulations are non-competitive, and even those which are competitive are not used mainly for the competitive element. Of the four examples given in Chapter 2, DART AVIATION LTD has no sociological or ethical content. The aim is to make money, to 'win', to produce good aircraft, to operate efficiently, to work out the effects of expenditure on advertising and the effects of pricing policy, to co-operate with other members of the team, etc.

However, to say that it lacks ethical or non-materialistic values is not to say that it is immoral, perverse, or likely to corrupt the young. If the teacher is afraid that the use of such simulations may have a harmful effect, it is quite easy to balance them with simulations dealing with ethical values. The simulation TENEMENT can arouse strong feelings of sympathy, denunciations of injustice, and pleas for a caring society.

SPACE CRASH might be regarded as competition against the environment. The aim is to survive and the environment is hostile. If SPACE CRASH has an ethical content, it is minimal. STARPOWER is ambiguous. Ostensibly it is highly materialistic in the trading sessions, yet when it comes to the question of changing the rules, ethical considerations sweep in.

This is perhaps a clue to ethics in simulations. If the 'rules', the 'content' or the 'guidelines' exclude ethics, everyone understands. Ethics are not disregarded — they are not there to be disregarded. But if ethical considerations are part of the simulation, they become matters of concern.

Finally, much of the criticism mentions the word 'games'. As Ravensdale says, game participants are 'competitors'. But in competitive simulations the motive is educational as well as creating a desire to 'win'. If the object was simply to generate intense competition, it would be better to use games rather than simulations. Even highly competitive simulations have a case-study element which helps to distinguish them from games and gaming.

Chapter 6

The way ahead

Expectations

If educational techniques were used according to their merits,
simulations would be commonplace in schools and colleges. The
pioneers of the 1960s hoped for, and perhaps expected, a fairly
dramatic surge forward. But although the 1970s saw a large
increase in the number of different simulations published,
particularly in the United States, the number of titles was deceptive.
Games, exercises and even puzzles were often included in the totals,
and most simulations were the product of small publishing houses
using short print runs, with sales measured in tens or hundreds
rather than thousands. Consequently, prices tended to be high to
match the relatively expensive cost of printing small quantities.

Particularly disappointing was the resistance to simulations in
secondary schools, where arguably the need for their use was the
greatest. The Bullock Report (1975), *A Language for Life,* said:

> As a consumer, a worker, a voter, a member of his community, each
> person has pressing reasons for being able to evaluate the words of
> others. He has equally pressing reasons for making his own voice
> heard. Too many people lack the ability to do either with confidence.
> Too many are unable to speak articulately in any context which might
> test their security. The result can be acquiescence, apathy, or a
> dependence upon entrenched and unexamined prejudices. In recent
> years many schools have gone a very long way to asserting this aspect
> of education as one of their most important responsibilities. But there
> is still a great deal to be done. A priority objective for *all* schools is
> a commitment to the speech needs of their pupils and a serious study
> of the role of oral language in learning.

Although the Bullock Report indicates a clear need which
simulations could help to fill, it does not seem to understand what
a simulation is. It appears to follow the dictionary definition —
feigning, pretending, mimicry, etc. In one passage it criticizes the
artificiality of 'the weekly period devoted on (*sic*) lecturettes,
"formal" debates, and mock interviews', but goes on to praise

a group of children who were giving talks while handling materials. It says:

> It is important to note that the children were being themselves, not obliged to play roles, or to see the situation as different from what it was in reality, or to imagine anything that was not the case. There is often great excitement and high motivation in simulation, role-playing, and constructing imaginary situations, but activities based on actualities ought not to be neglected.

The word simulation is not mentioned in either the glossary or the index of the Bullock Report. The reference to 'great excitement' and 'high motivation' is double-edged praise, almost implying that what it gains in excitement may not compensate for what it allegedly lacks in hard reality. After all, a row of fruit machines in the classroom might generate even higher excitement and motivation.

The Bullock Report is one of countless examples of simulations being damned by the dictionary. *The Times Educational Supplement* ran an article on simulations (5 September 1975) under the title 'Mimicry and make-believe — in pursuit of the real thing'. To label something as mimicry, make-believe, artificiality and unreality is hardly likely to create a wave of enthusiasm in educational circles.

There were signs of some enlightenment in the 1970s, particularly from the BBC which produced a series of ten radio programmes on simulations. The accompanying booklet, edited by Longley (1972), began:

> Simulation, role-play and games are not new aids to teaching but it is only recently that the techniques have been used widely in education and training. The military have their strategy games, business schools their management simulations, colleges of education their in-tray exercises and primary schools their 'shops'. However, in secondary schools and further education these techniques have still to be exploited.

The 1970s were a period of unsteady progress in which simulation development was fairly successful in some areas, while highly disappointing in others.

The 1980s — several oil crises on — are hardly likely to see a massive breakthrough. With educational establishments short of money and simulations still dogged by misleading terminology, there are good reasons for being cautiously pessimistic.

However, the analogy of 'breakthrough' suggests that it is simulations which are the moving objects trying to penetrate the defences of educational establishments. This is not necessarily the case. The defenders can foray out and capture simulations if they suddenly appear to meet a need. To continue the military analogy, different things happen on different parts of the battlefield.

The following three areas — secondary schools, business education, and the teaching of English as a foreign language — are chosen not because they are typical but because they are different from each other. Perhaps the best hope for simulations in the 1980s is that development and publicity in one area will help to spread the technique to other areas.

Secondary schools

Secondary schools resist simulations to an extent which does not seem to apply in other educational establishments. There are, of course, notable exceptions, and the number of simulations used in secondary schools is on the increase. But the general picture is one of 'don't know, don't want to know, too busy'.

By far the most authoritative account of what has been happening in secondary schools during the 1960s and 1970s is given in *Aspects of Secondary Education in England* (1979). This is a report of an investigation by HM Inspectors covering 384 schools, about 10 per cent of the total. It is one of the most thorough investigations ever carried out in Britain. What they have to say reveals why simulations, together with other forms of spoken language, have been having such a hard time.

The report found that there were too many examinations and too many optional courses. Although most schools were hard-working and orderly communities (lack of discipline was a source of anxiety in only 27 of the 384 schools) and there was no lack of attention to basic skills, there had been a profligate increase in written work, a lot of it just copying. The inspectors found one fifth-form group who had written from dictation 23,000 words of plot from *Far from the Madding Crowd.*

An able boy in the second term of his fourth year wrote 6000 words in English, 10,000 in history, 6000 in French, 2300 in geography, and 10,000 in music — most of the writing being copied notes. It is an example, say the inspectors, which could too easily be replicated, with only slight variations.

With far more pupils involved in public examinations and with some teachers over-quick to hand out writing to keep a class quiet, there was a common pattern of massive notes, essays, answering of examination questions, tests and exercises which crowded out opportunities for the pupils to experience, and more particularly use, language.

The inspectors found that the recommendations of the Bullock Committee (1975) had little effect, and they doubted whether the phrase advocating 'language across the curriculum' had even been properly understood. Many teachers assumed that the only features of language which were important were those relating to 'correctness'. That function, thought the teachers, was the responsibility of the English department.

Language used by pupils to show that they had learned something — replies to teachers' questions — was one thing, said the inspectors, but language used in achieving learning was a very different thing and was both an important need and a diminishing element in secondary schools.

Various examples of good language usage listed in the report included pupils in a school's rural studies department who planned the use of enclosures and feeding troughs for farmstock. In other words, this was a simulation, although the inspectors do not use the term.

As a result of these findings, the inspectors called for a professional revaluation of the 'largely unchallenged pre-eminence of writing'. The report, published as a government White Paper, is a clear signpost for the 1980s. It is authoritative, consistent, and has been widely welcomed by the teaching profession. It also marks an end to the concept of the inspector as a sort of policeman. The inspectors see the report as the start of a new era of responsible self-evaluation within schools, with the report itself being a sort of do-it-yourself inspection kit.

Because the report clearly identifies a major fault — too much writing and copying — it seems plausible that the decline in spoken language will be halted and possibly reversed in the 1980s. With the evidence supplied by the inspectors as ammunition, more voices are likely to be raised in favour of projects, simulations, role-play exercises, case studies, and the like. Schools may chase simulations, instead of the other way round. It may be that teachers who make a practice of using simulations will also use them on occasions for testing and assessing their pupils.

Business

Business has never been submerged by the written word. On the shop-floor, in the boardroom and in the market, the spoken word has had an important place.

Free from the strange inhibitions which affected the secondary schools, businesses in both the private and public sectors have been using simulations to an ever-increasing extent. Sometimes they are tailor-made by the organization's own staff training department, sometimes outside designers are called in, and sometimes the simulations are bought ready-made.

For non-managerial staff, the simulations are usually replicas of actual operations the organization is involved in, in order to familiarize the staff with the materials, problems and routines. The aim is training rather than education.

For managerial staff a large proportion of the simulations are related to jobs and the procedures and problem-solving of the company's management. But there is a tendency to widen the horizon when it comes to recruitment and selection for possible promotion. Simulations are now taking their place along with psychological tests, problem-solving and decision-making exercises which may have little direct connection with the specific goods or services produced by the organization. Moreover, simulations are spreading outwards from business into education.

Most large companies and state corporations have education departments which are taking an increasing interest in what goes on in schools, colleges and universities. Some of these organizations — large banks, oil companies — produce materials specifically for use in education. The prime example, of course, is the activity of the BBC in education. It is now no longer unusual for a large organization to approach an author and ask for a simulation which conveys some aspect of its products or services for use in education at secondary school level or above. There is good reason to believe that this practice will continue and probably increase. If an organization has its product associated with a pleasurable and useful activity in the classroom, it is likely to regard this as a desirable development for reasons of self-interest, altruism, or both.

An alternative form of promotion is sponsorship which is unconnected with the organization's products. Townsend's *Five Simple Business Games,* of which DART AVIATION LTD is one, was sponsored not by an aircraft company, but by the British Oxygen Company. Since the arts, sport and games are already being sponsored by firms to an ever-increasing extent, simulations could well become candidates for sponsoring by commercial organizations.

Whether this is a good thing is debatable. Publishers and authors of commercially produced simulations might not be particularly pleased about sponsored cut-price simulations, considering them unfair competition. The same, of course, could be said of subsidized simulations published by educational bodies — universities, learning materials centres, etc. On the other hand, commercial publishers have been slow to enter the simulation field.

Some educators may not be happy about the intrusion of commercial interests into the classroom, feeling that their pupils are being insidiously 'got at'. A publisher's name on their own books is one thing but a sponsored book, project, simulation, lecture, is another. Other teachers, however, are only too pleased to grab whatever useful materials come along and are not deterred by sponsorship.

It could be argued in favour of sponsorship — or subsidized simulations — that they promote the simulation concept, thus widening the market; and the wider the market the less need for short-run, high-cost simulations.

In any event, business-motivated simulations are expanding into education, and this seems likely to continue.

English language

English is the language of international transport, commerce, much of education, and diplomacy. Among the nations of the world, English is either at the top of the language league or is moving up it.

In the teaching of English as a foreign language the use of simulations is certain to increase in the 1980s. It is an aspect of education in which teachers have sought the simulations rather than simulations seeking teachers. Some simulations designed for quite different students are being used for teaching language skills to foreign students. The work of the British Council and the Council of Europe has already publicized the point that simulations, together with role-play, are useful things.

Foreign students of English are often adults. In this market, therefore, as in the market for business simulations, there is more money about, particularly for educationally valuable cost-effective materials.

Nor is it just a question of learning English. Teachers like materials which generate enthusiasm for their use and are a pleasure to introduce, not a chore. Functionalism is not just a philosophy about using English as a tool to do a job (as distinct from doing an exercise to please the teacher); it has a human aspect for both teacher and student.

Some of the credit for this state of affairs goes to the magazine

Modern English Teacher which has a world-wide circulation and is full of practical ideas about suitable materials and techniques.

The BBC is also a major progressive force in learning English, both through its publications and through its projects English by Radio and English by Television. Nor is it unusual when teachers of English are being interviewed to hear a question like, 'Do you use simulations?'

The United States, which is the home of most simulations, appears to be slow in using them for teaching English as a foreign or second language. Canada, on the other hand, seems more progressive in adopting functionalism than its big neighbour, which may be partly due to British influence.

In many foreign countries there is a reluctance to depart from rote learning but things are gradually changing, and certainly there is enormous scope for the use of simulations in EFL teaching.

If adults start to use simulations — in whatever field — it might not be long before they start saying 'If it is good enough for me, it is good enough for my children.' Educational ideas are not pillars of marble. They flow around from one aspect of education to another, and influence and are influenced by the thinking of parliaments, parents and inspectors. Simulations in the 1980s may not achieve any dramatic breakthrough, but one does not have to be an optimist to forecast expansion.

Simulations in national examinations

In February 1980 the Department of Education and Science announced its decision to merge the two big secondary school examinations — the GCE and the CSE. Long negotiations on the contents, standards and arrangements are envisaged. The new system is not expected to be ready until 1986. Mode Three of the CSE examination, which includes an assessment of the pupils' ability to talk effectively, would be retained but within a tighter framework.

One problem facing the examination boards and the Inspectorate is to reconcile the need for subjective tests with the practical problem of national standards. Who are to be the assessors? What training would be required? Are adequate resources available? These are the sort of questions which will be discussed within the educational establishment, and outside it.

The present position, as mentioned earlier, is that classroom teachers can use simulations — together with informal drama, interviews or debates — as part of the assessment for Mode Three. The use of simulations in national examinations raises two main issues: practicality and desirability.

On the question of desirability, the Bullock Report gives strong, although indirect, support for using simulations as a means of assessing the use of oral language. As already seen, the Bullock Committee appears to accept the dictionary definition of the word simulation. But it appears to include the concept of simulation as part of informal drama. In Part 4 of the Bullock Report, in the section on oral language, the use of informal drama as a form of subjective assessment is welcomed. It says:

> We do, however, have one important reservation. In devising the Mode Three syllabuses a number of teachers are placing heavy emphasis on the 'history of theatre', with an undue weight on the learning of facts completely detached from any practical work. It would be unfortunate if a quest for 'academic respectability' for the subject led to an increase in syllabuses of this type. In our view the greatest value to be gained from the development of examination work would be in expansion of the kind of complementary activity described in paragraph 10.33.

Paragraph 10.33 of the Bullock Report is exceptionally interesting, since it stresses the importance of combining the written with the spoken word.

> There are countless occasions when written words — not just those in a play — are illuminated by being placed in a real context, which drama can help to realize. In its turn improvisation can be enriched by the written word. This does not mean that the written word should be imposed upon the activity. It means that it can provide the origin and stimulus, the 'story', the 'situational contexts' for the work in improvisation. In other words, improvisation can be initiated or given substance by literature, for here may be found the characters, relationships and situations for imaginative work in improvised drama. What is so often lacking in improvisation is stimulus and subject matter of quality, and literature is an unequalled source of this. We have seen many improvised scenes in which the spontaneous language produced by the children was of limited range and interest, often rapidly degenerating into a trivial slanging match. Unless the stimulus of good writing (whether prose, verse, or drama) is offered from time to time the improvised dialogue will too often derive weakly from playground scraps and casual chats. The extending value of improvisation will then be lost. Nevertheless, quite apart from its other qualities, it is improvisation, involving the complicated relationships between the written and the spoken word, which seems to us to have particular value for language development.

Although this passage stresses the idea of improvisation, it also stresses the need for a real context, a real situation, and something written down. Thus the improvisation should be within a written framework. Although the passage is about informal drama, it seems to cover a good many simulations as well. This is within the context of recommendations of material suitable for examination assessment as well as for learning.

Turning from the question of desirability to that of practicality, various questions arise — selection of suitable simulations, standards of assessment, training of assessors, and training of teachers to become effective controllers. Clearly, test simulations cannot be selected from a list with a pin. Each participant should have reasonable opportunities to talk. Part-time roles, passive roles, and non-speaking roles should be avoided. The format of a suitable simulation could involve a public meeting, committee meeting, debate or interview. But it should avoid having someone in the role of, say, secretary who has nothing to do but take notes. There is no reason why suitable simulations should not be tailor-made for the examinations, just as examination questions are tailor-made. Standards of assessment might be a problem, but this is a general problem affecting all forms of subjective assessment. If informal drama can be assessed, so can talk in simulations. If anything, the assessment could be easier to standardize. Moreover, it is not the simulation materials that are being assessed but the oral language of the participants, and in essence this is no different from assessing the oral language in a set speech or formal debate. In a simulation the context is real rather than artificial, but the problem of assessing the language is no more difficult than in a formal speech and may be easier and have a more reliable result.

The big question, however, is not likely to concern the form in which oral language is examined but who should assess it. Should teachers assess their own pupils, or should the assessment be done by outside assessors?

The training of assessors is a separate problem which depends on the answer to the previous question — who should be the assessors? But it is a mistake to assume that, because simulations are a relatively new form of learning in secondary schools, there must necessarily be a massive training programme for simulation assessors. It is not simulations that are being assessed; it is the quality of the talk of the participants. Moreover, a distinction must be made between the assessment of the language and the controlling of the simulation. The assessor should not be the controller, any more than the assessor of an informal drama should be the person who introduces it to the pupils. While, by the nature of things, assessors of simulations are likely to be good controllers, and assessors of informal drama are likely to be good at introducing informal dramas, they do not have to be.

113

The training of teachers to be éffective controllers is not a major problem. As suggested in this book, it is largely a question of avoiding certain misconceptions, using commonsense and gaining experience. If simulations were included in national assessments, teachers would introduce simulations to help their pupils, and this would help teachers to become experienced controllers. Educational bodies already hold courses and workshops on simulation techniques just as they do on informal drama, and doubtless their number would increase with the impetus of examinations.

Simulation development

There are many things which authors, publishers, teachers and the education authorities can do to improve simulation development. Some improvements are inevitable. Increased practice will bring greater experience, awareness, skills and confidence. Publicity and information about simulations will help to reduce misunderstanding. However, authors should make positive efforts to provide adequate information and advice in the controller's notes. For example, it is not unusual to come across a teacher who has tried out one simulation only. When asked why he or she has stopped at one simulation, the reply is something like this: 'The pupils seemed to be having a lot of fun, at least some of them did, which was quite useful in its way. Early on they got into a good discussion, but there was a bit of argument about the materials and about what was what. I'd say that, on the whole, they enjoyed it, but I'm very doubtful whether they really learnt anything. They didn't seem terribly keen on doing another one.'

This leaves a row of question marks. Did the teacher expect to be able to measure benefits? Did 'learnt anything' refer only to facts, or did it take skills into account? Was the de-briefing thorough, or was it a perfunctory 'What did you think of it?', 'We thought it was all right.' What was the problem about materials? Why was there a good discussion early on, but not later? What was meant by 'fun' and 'enjoyed' — shrieks of laughter or quiet appreciation? Was it an informal drama or a simulation? Was it a genuine simulation or a runaway simulation?

The fault may be with the author for failing to provide adequate controller's notes. Unfortunately, simulation after simulation is published with only the briefest description and advice. The objectives are written large, there are a few suggested questions for the de-briefing which are related to the objectives, and the mechanics are explained briefly. But rarely is there any explanation of what a simulation is, what should be avoided, and what are the possible pitfalls. There are exceptions, of course. Garry Shirts' BAFA BAFA, for example, has very clear notes, with helpful suggestions about actual arrangements in the classroom, together with a discussion of the sociological and philosophical content.

Adequate controller's notes will not transform a poor simulation into a good one, but at least they should be educative. It would also help if the controller's notes explained that simulations were events. This at least would be a warning to non-users who picked up the materials intending to inspect them in the same way that they inspect a book.

Almost certainly both the media and the educational authorities will do more in the 1980s to inform and enlighten teachers — and themselves — about the nature and potentiality of simulations. There should be more courses, conferences and workshops where teachers can participate in simulations. Improvements in library facilities and electronic developments such as ERIC (an American computerized clearinghouse on information resources) will enable documents and articles on simulations to be more readily available on an international scale. Examinations can be a key issue. If simulations are included among the tools of assessment at an official level, a big step forward will have been taken.

The other feature of simulation use which has been neglected is in the field of consumer reports. It is sometimes the case that certain simulations always tend to stick at specific points, although otherwise they are very useful. Occasional articles in simulation literature try to remedy the defects. It has been known for enthusiasts to pass on the simulation accompanied by a helpful article written by a previous user. This sort of practical article has a general as well as a specific use, since it sharpens the critical faculties of the user. It makes him more aware of what is happening, and better equipped to deal with it.

Compared with the 1960s, there are many more good simulations available and the 1980s will see further improvements. Useful ideas and techniques will be noted and copied. Probably the research trend will be away from trying to justify high and noble objectives, and will be towards observing what is actually going on and suggesting ways of improving it. The demand from various educational organizations for simulations as a technique for encouraging language, decision-making and communication skills is bound to increase, together with the development of specific simulations to meet specific local needs. Prediction is always hazardous, but the likelihood is that simulations will make uneven but significant progress in the next decade.

References

Abt, C C (1968) Games for learning. In Boocock, S S and Schild, E O (eds) *Simulation Games in Learning* Sage: California

Bloomer, J (1973) What have simulation and gaming got to do with programmed learning and educational technology? *Programmed Learning & Educational Technology* **10** 4

Bloomer, J (1974) Outsider: pitfalls and payoffs of simulation gaming *SAGSET News* **4** 3

Boardman, R (1969) The theory and practice of educational simulation *Educational Research* **11** 3: 179

Bullock, A (1975) *A Language for Life* Report of the Committee of Inquiry appointed by the Secretary of State for Education and Science under the Chairmanship of Sir Alan Bullock. HMSO: London

Coleman, B Videorecording and STARPOWER *SAGSET Journal* **3** 1. Also published in Megarry, J (1977) (ed) *Aspects of Simulation and Gaming* Kogan Page: London

Davison, A and Gordon, P (1978) *Games and Simulations in Action* Woburn Press: London

Elgood, C (1976) *Handbook of Management Games* Gower Press: Farnborough

Garvey, M D (1971) Simulation: a catalogue of judgements, findings and hunches. In Tansey, P J (ed) *Educational Aspects of Simulations* McGraw-Hill: London

Guetzkow, H *et al* (1963) *Simulations in International Relations* Prentice-Hall: Englewood Cliffs

H M Inspectors of Schools (1979) *Aspects of Secondary Education in England*. A survey by H M Inspectors of Schools. HMSO: London

Jones, K (1974a) *Nine Graded Simulations* (SURVIVAL, FRONT PAGE, RADIO COVINGHAM, PROPERTY TRIAL, APPOINTMENTS BOARD, THE DOLPHIN PROJECT, AIRPORT CONTROVERSY, THE AZIM CRISIS, ACTION FOR LIBEL). First published by ILEA, now published by Management Games Ltd: Bedford

Jones, K (1974b) Review of TENEMENT *Games and Puzzles*, June, No 25

Jones, K (1980a) Communication, language and realsits. In Race, P and Brook, D (eds) *Perspectives on Academic Gaming & Simulation 5* Kogan Page: London

Jones, K (1980b) SPACE CRASH (also ROCK ISLAND TRANSPORT, OUTERWORLD TRADE and IS GOD THERE?) Management Games Ltd: Bedford

Kerr, J Y K (1977) Games and simulations in English-language teaching. In *Games, Simulations and Role-playing* British Council: London

Longley, C (1972) (ed) *Games and Simulations* BBC: London (out of print)

Masterman, L (1980) *Teaching about Television* Macmillan: London

Ravensdale, T (1978) The dangers of competition *SAGSET Journal* **8** 3

Shelter (1972) TENEMENT Shelter: London

Shirts, R G (1977) BAFA BAFA Simile II: La Jolla, California. Also available from Management Games Ltd: Bedford

Shirts, R G (1970) Games people play *Saturday Review,* 16 May

Shirts, R G (1969) STARPOWER Simile II: La Jolla, California. Also available from Management Games Ltd: Bedford

Tansey, P J and Unwin, D (1968) Simulation and academic gaming: highly motivational teaching techniques. In Dunn, W R and Holroyd, C (eds) *Aspects of Educational Technology II* Methuen: London

Taylor, J L and Carter, K R (1970) A decade of instructional simulation in urban and regional studies. In Armstrong, R H R and Taylor, J L (eds) *Instructional Simulation Systems in Higher Education* Cambridge Institute of Education: Cambridge

Townsend, C (1978) *Five Simple Business Games* (GORGEOUS GATEAUX LTD, FRESH OVEN PIES LTD, DART AVIATION LTD, THE ISLAND GAME, THE REPUBLIC GAME) CRAC/Hobsons Press: Cambridge

Twelker, P A and Layden, K (1973) A basic reference shelf on simulation and gaming. In Zuckerman, D W and Horn, R E (eds) *The Guide to Simulations/ Games for Education and Training* Information Resources Inc: Lexington

Twelker, P A (1977) Some reflections on the innovation of simulation and gaming. In Megarry, J (ed) *Aspects of Simulation and Gaming* Kogan Page: London

Zuckerman, D W and Horn, R E (1973) (eds) *The Guide to Simulations/ Games for Education and Training* Information Resources Inc: Lexington

Bibliography

Abt, C C (1970) *Serious Games* Viking Press: New York

Armstrong, R H R and Taylor, J L (1970) (eds) *Instructional Simulation Systems in Higher Education* Cambridge Institute of Education: Cambridge

Armstrong, R H R and Taylor, J L (1971) (eds) *Feedback on Instructional Simulation Systems* Cambridge Institute of Education: Cambridge

Banks, M H, Groom, A J R and Oppenheim, A N (1970) Gaming, simulation and the study of international relations in British universities. In Armstrong, R H R and Taylor, J L (eds) *Instructional Simulation Systems in Higher Education* Cambridge Institute of Education: Cambridge

Bloom, B S *et al* (1956) *Taxonomy of Educational Objectives. Handbook 1: Cognitive Domain* Longmans: London

Bloomer, J (1975) Paradigms of evaluation *SAGSET Journal* **5** 1: 36-7

Boocock, S S and Schild, E O (1968) (eds) *Simulation Games in Learning* Sage: California

Cherryholmes, C H (1966) Some current research on effectiveness of educational simulations: implications for alternative strategies *American Behavioral Scientist* **10** 2: 4-7

Clarke, M (1978) *Simulations in the Study of International Relations* G W and A Hesketh: Ormskirk

Cruickshank, D R *et al* (1979) The state of the art of simulation in teacher education *Simulation/Games for Learning* **9** 2: 72-82

Duke, R D (1974) *Gaming: The Future's Language* Wiley: New York

Gibbs, G I (1974) (ed) *Handbook of Games and Simulation Exercises* E and F N Spon: London

Hauser, R E (1977) The ethics of gaming *SAGSET Journal* **7** 4

Jones, K (1973) Simulation and communication skills in secondary schools *Educational Research* **15** 2

Jones, K (1977) Simulations for the sake of talking. In Megarry, J (ed) *Aspects of Simulation and Gaming* Kogan Page: London

MacDonald-Ross, M (1973) Behavioural objectives: a critical review *Instructional Science* **2** 1

Megarry, J (1976) Ten further 'mistakes' made by simulation and game designers *SAGSET Journal* **6** 3

Megarry, J (1977) (ed) *Aspects of Simulation and Gaming* Kogan Page: London

Merriman, N (1975) Legal aspects of playing other people's games *SAGSET Journal* **5** 3

Shirts, R G (1975) Ten mistakes commonly made by persons designing educational simulations and games *SAGSET Journal* **5** 4: 147-50

Stadsklev, R (1975) (ed) *Handbook of Simulation and Gaming in Social Education* Institute of Higher Educational Research and Services: University of Alabama

Stenhouse, L (1975) *Introduction to Curriculum Research and Development* Heinemann: London

Sykes, P (1978) New moves in business games *SAGSET Journal* **8** 1

Tansey, P J (1971) (ed) *Educational Aspects of Simulation* McGraw-Hill: London

Tansey, P J and Unwin, D (1969) *Simulation and Gaming in Education* Methuen: London

Tawney, D (1976) *Curriculum Evaluation Today: Trends and Applications* Macmillan: London

Taylor, J L and Walford, R (1978) *Learning and the Simulation Game* Open University Press: Milton Keynes

Van Ments, M (1978) Role-playing: playing a part or a mirror to meaning? *SAGSET Journal* **8** 3: 83-92

Walford, R A (1972) *Games in Geography* Longmans: London

Wilson, A (1969) *War Gaming* Penguin: London

Societies and journals

ISAGA: International Simulation and Gaming Association. Information can be obtained from Dinah Goldberg, University of Geneva, FPSE, 24 rue Général Dufour, CH-1211, Geneva 4, Switzerland.

NASAGA: North American Simulation and Gaming Association is the American equivalent of SAGSET. Information can be obtained from RWT Nichols, Treasurer, NASAGA, Box 100, Westminster College, New Wilmington, Pa 16142.

SAGSET: The Society for Academic Gaming and Simulation in Education and Training. SAGSET is a voluntary professional society. Its aims are to encourage and develop the use of simulation and gaming techniques in all applications in education and training. Most of the members are British educationalists. More information can be obtained from The Secretary, SAGSET, Centre for Extension Studies, University of Technology, Loughborough, Leics LE11 3TU.

The quarterly journal of SAGSET is *Simulation/games for learning,* formerly *SAGSET Journal.*

The proceedings of SAGSET conferences from 1975 are published in:
Perspectives on Academic Gaming and Simulation 1 & 2 (1975 and 1976)
Perspectives on Academic Gaming and Simulation 3 (1977)
Perspectives on Academic Gaming and Simulation 4 (1978)
Perspectives on Academic Gaming and Simulation 5 (1980)
All are published by Kogan Page.

Games and Puzzles: A British quarterly about recreational games and puzzles with some items of educational interest. Willow House Press, 26 Parkway, London NW1.

Journal of Experiential Learning and Simulation: Academic with an inclination towards business simulations. Elsevier North Holland Inc, 52 Vanderbilt Avenue, New York, NY 10017, USA.

Modern English Teacher: A quarterly magazine of practical suggestions including simulations, games and role-play for teaching English as a foreign language. From Modern English Publications Ltd, 33 Shaftesbury Avenue, London W1V 7DD.

Simgames: An informal Canadian quarterly available from Simgames, Champlain Regional College, Lennoxville Campus, Lennoxville, Quebec, Canada, J1M 2A1.

Simulation and Games: An academic US-based quarterly available from Sage Publications, 275 S Beverly Drive, Beverly Hills, Ca 90212 or from 28 Banner Street, London EC1.

ERIC: At many educational institutions the complete ERIC microfiche file is available in the library, and a particular document can be located by using the ED number and can be ordered at the stated prices, plus postage, from ERIC Document Reproduction Service, Box 190, Arlington, Va 22210. Only documents with ED numbers are available in this way.

Resource-Finder Service: Many simulations, games and books are available through MGL Resource-Finder Service, Management Games Ltd, 11 Woburn Street, Ampthill, Bedford, MK45 2HP.

Index